THANK
EVOLUTION
FOR
GOD

The Roles of Nature and God In Evolution

Revised

LOUIS W. PERRY

This Revision expands the breadth of the review of Christian authors who address the evolution-religion conflicts with the addition of a sixth book, *What Makes Christianity So Great?* by Dinesh D'Souza. The author is not a scientist, but a writer who inserts his strong Christian and conservative political ideologies into debates and books on the interactions of Christianity with science, including Darwinian evolution and the government.

Contents

To my mother, Zada Arline Perry
for her loving guidance and encouragement.

Acknowledgments

This book is an outgrowth of discussions with students over a series of seminars on the "conflicts of religion with science and democracy" at the University of California San Diego. Many thanks are given to the students of many different religious persuasions for their contributions to the lively and insightful questions and discussions.

Thanks to Provosts Tom Bond, Barbara Sawrey, Dan Wulbert, and Don Wayne of Revelle College for supporting the seminars over several years. Also thanks to Professor Francisco Ayala at the University of California Irvine, Professors Patricia Churchland, Paul Churchland, Russell Doolittle, and Alfred Zettner at the University of California San Diego for their biological and neuro-scientific insights. Special thanks to Professors Scott Hestevold, Norvin Richards, and Richard Richards at the University of Alabama for their most helpful discussions on the philosophy of evolution. Additionally thanks to the members of the Bible study group of the University Christian Church for the views of progressive Christians and science.

The words, thoughts, and errors are mine alone and not those of the individuals or the universities noted.

Prologue

This is a photo of a T-shirt from the Philosophy Department of the University of Alabama honoring Socrates as a teacher to students, citizens, and governments with the inscription:

2,500 years and still corrupting the youth.

Socrates, philosopher, teacher, war hero, and inquisitive citizen, was one of the most famous Greeks of his time. As a philosopher Socrates used questioning to search for knowledge and even to search for the nature of knowledge. This method made him one of the early giants of philosophy, but he got himself and his students into trouble by agitating the city government of Athens over many issues. He was a general pain in the neck to the authorities and others living their lives with conventional wisdom. When he would not change his ways, Socrates was charged in 399 BCE with a religious crime by the officials of Athens, who were embarrassed by his questioning of their conventional wisdom. His crime:

> *... failing to acknowledge the gods that the city acknowledges and introducing new deities.*

He was found guilty and sentenced to death by the theo-cratic Athenian government, which had the power to define political and religious correctness and when necessary use that power to convict anybody of anything. Socrates died by drinking poison given to him by the state. Since that time, questioning religious dogma by philosophers, scientists, and lay people has continued, and at times it has also been hazardous to their health. It is thought that Socrates would approve the questioning of the conventional religious dogma currently causing conflicts between religion, science, and the state.

Departmental sponsor: *Philosophy*

Introduction

Toward the end of the university's spring academic quarter last year, one of my students asked if I had read the book *Thank God for Evolution* by Michael Dowd, a Christian minister who declared there was no conflict between science and religion, for God had created both religion and science, including evolution. It was an interesting thought, for the course I was teaching centered on understanding the conflicts of religion, science, and democracy. I had not read the book, so I demurred and promised to read and comment on it.

As I was reading the book, it became clear to me that Dowd's message was written from his supernatural God's perspective, and accordingly raises many questions about his understanding and appreciation of the science of evolution. From the perspective of Nature, he simply failed to honor the science of the natural world. Dowd rejected Darwin's Nature-centric theory of natural selection and substituted a God-centric theotheory he invented so that he could declare *thank God for evolution.*

His approach opens a hundred questions, the first being: Why does Dowd inject God into Darwin's Theory when he knows that scientists argue that it works quite well without any involvement by a supernatural god? But Dowd's view is not alone, so to put his book in context with the views of other Christian authors several other Christian books on evolution were considered. My comments grew to many pages and this book is the result.

Dowd's title, *Thank God for Evolution,* is an expression of his Christian belief that God is the cause of evolution. But Darwin's Theory cannot be rejected so easily, for over the last 150 years it has been tested over and over and seems to work quite well without the involvement of the Christian God or any god. But Dowd and many other Christians refuse to accept the vast amount of data supporting the theory and insist their God be involved. Fortunately, there are some Christians, the Progressives, who, along with most Naturalists, do look at the scientific data and accept Darwin without God's involvement.

Dowd's supernatural[1] explanation of man's evolution is a common one among Liberal[2] Christians. They mix religion and science together and invent theotheories that puts God in charge of evolution. The mixing then allows them to declare there are no conflicts between science and religion. Religious theotheories of evolution mixed with Nature's processes[3] produce theotheories that are fine for use in theological sermons to believers who hold that God can do anything, but not so fine for the scientific community, who hold that mixing religion and science only produces a religious muddle and scientific nonsense.

Dowd's title implies a hypothesis that places Nature subservient to God, an acceptable Christian belief but an unacceptable position for Naturalists. When Dowd's title is viewed from Nature's perspective, one imagines an antipodean view based on Nature which allows for natural evolutionary processes to produce man and his brain with an evolved moral sense and mental capabilities for developing religious narratives and god concepts. The scientific community[4] accepts Darwin's Theory as settled science, for it has accumulated mountains of supporting data over the 150 years since the publication of *On the Origin of the Species by Means of Natural Selection*, Darwin's seminal book on evolution. So the science we know today offers an alternative: we can *thank evolution for God* as a compelling scientific hypothesis emerging from our current understanding of man's evolution. This book offers an approach to understanding evolution that retains the integrity of natural science, governance, and supernatural religions.

Questioning the authority of God in the natural world (He is not questioned in the supernatural world) on the subject of man's evolution is very likely to step on many toes and put the author into the situation Socrates found himself, questioning everything, including the gods, and getting into trouble. But I hope the reader will hold off prescribing the hemlock for the author to drink until after reading the full arguments about *who to thank for evolution—God or Nature*

Approach

It is clear from Dowd's book, and other similar books on man's evolution by many Christian authors that there are fundamental differences between scientists and Christians in their understanding and acceptance of Darwin's Theory of natural selection, what the Darwinian Revolution is all about, what evolutionary biologists are telling us about living organisms, and what neuroscientists are telling us about man's evolved mental capabilities and moral sense. It quickly becomes apparent that Dowd in his discussion of evolution does not wish to be constrained by the laws of Nature as are scientists, but guided by his beliefs on what he thinks his supernatural God would say about natural science. His book, in essence, is a sermon on evolution using a theotheory he invented as a Christian. This is fine for believers, but not so fine for Naturalists and scientists who shudder when theotheories invented in the Christian supernatural world[5] are mixed with science theories in the natural world.

Dowd claims that his book is useful for all people, but since it is about the supernatural world of his Christian God and includes miracles, its usefulness is primarily limited to the Christian community. So at best his God-centric answer is acceptable to Christians, about one-seventh of the world's population, and unacceptable to the six-sevenths who are not Christian. Is there not an understanding of Darwinian evolution that is acceptable to all peoples of all religions?

Dowd's invented theotheory of evolution is neither good Christian theology nor good Darwinian science, for it rejects both God's description of man's creation in Genesis and Nature's Darwinian evolution and replaces them with an invented theotheory of evolution, Sacred Innovative Design (my name). In effect, Dowd is contending that his God got it wrong in Genesis by not including Darwinian evolution and that Darwin got it wrong by not including God in his theory, so it was necessary for Dowd to propose a supernatural theotheory to fix God's errant Scriptures and Darwin's inadequate theory. With his invented Sacred Innovative Design theotheory, Dowd then declares he has brought harmony for Christians to the conflicts over man's evolution and, as his title states, that all can rejoice and *thank God for evolution.*

Is this only Dowd's view? What about other Christian authors who have written on evolution? To find out how Dowd's proposed theotheory compares with those of other Christian authors, I have included five additional books in the study: *The Language of God: A Scientist Presents Evidence for Belief* by Francis Collins, a biologist/administrator in the government, *Coming to Peace with Science: Bridging the Worlds Between Faith and Biology* by Darrel Falk, a professor of biology at a Christian college; *Saving Darwin: How to Be a Christian and Believe in Evolution* by Karl Giberson, a professor of physics at a Christian college; *What's So Great about Christianity?* by Dinesh D'Souza, a conservative political writer; and *Darwin's Gift to Science and Religion* by Francisco Ayala[6], a professor of evolutionary biology at a public university. Additionally, I've included a Web site (BioLogos) established by Collins as an additional source of Liberal Christian views on evolution.

The thrust of the first five books by the Christian authors, including the one by Dowd, can be expressed by expanding Dowd's title:

Thank {the Christian} God for {God-directed Darwinian} Evolution.

Five of the Christian authors, like Dowd, also employ various invented Christian theotheories that mix science and religion together. In doing so, they consider Darwin's Theory fair game—that is, it can be rewritten and molded to an author's will, and never mind that in the process, natural science takes a beating. The result from a Christian view is that harmony and peace have been achieved between religion and science.

But the use of theotheories to bring peace by modifying Darwin's Theory comes with unintended consequences. For example, in the scriptures, God's creation of man and his gift of morals to him, whether one uses a metaphorical or literal view, are a onetime—a day or so—event, while in Nature, the evolution of man and his morals are a long-term process covering billions of steps over billions of years. This, of course, produces a conflict between the scriptures and science. The price Christians have to pay for using a supernatural solution is to suffer the slings and arrows from the scientific community for violating a well-tested natural science theory.

If they were clear statements by the authors that their books are Christian narratives employing metaphors, we could all agree that they are sermons and say fine. However, the approach they take mixes religion and science, violates natural science theories, and claims to be useful to all peoples. This only highlights the problem that theotheories, although acceptable in sermons for believers in their churches, are not acceptable to scientists or science classes in public schools.

The sixth Christian book, by Ayala, treats religion and science as separate and independent disciplines and does not mix the two, nor does it use an invented theotheory. In this case the discussion of

Darwin's Theory is good science, free of religious theotheories and acceptable in public schools. The importance of the independence of science and religion is critical and expanded upon later at length.

Naturalists reject Dowd's position of having a supernatural God involved in natural science and see it as a leftover from the old days when the Christian church exercised theocratic dominance over science and could rewrite scientific theories at their will. But time has moved on, and advances in science have produced scientific revolutions that have succeeded in peeling off many layers of old, supernatural Christian dogmas over the natural science that had formed during a thousand years of Christian theocratic governance. The past days of Christian theocratic authority over science are long gone, and accordingly, Christians must be reminded that Darwin's Theory of evolution, like all other natural theories, is rooted in Nature's world and needs no help from a supernatural God to work. Christians may do what they will with scientific theories for use in their sermons in churches, but these sermons have no validity in the natural world. To modify scientific theories, everyone—that includes scientists and Christians alike—must go through the scientific process step-by-step and get the approval of the scientific community; there are no short-cuts or appeals to God for help.

The approach taken is to ask the question, *Who should we thank for evolution—God or Nature?* to each Christian author and then compare their answers with one based on natural science. In developing the position for Nature, it was necessary to question Christian dogma and develop an understanding of the boundaries of the authority of God (religion), Nature (science) and the state (governance). No special effort was given to bring harmony, but care has been taken to see that religion and science are kept separate, thereby not violating natural science theories or supernatural religious beliefs.

The state, our democracy, was introduced as the third discipline in our discussions because it has bureaucratic authority over scientific information in countries and can impact science and public education. Our Constitution requires us not to have the state establish religions, a difficult rule for some overzealous Christians. The nation's public science classrooms and government institutions rely on the authority of the secular science community for science education. Christian theotheories on evolution are judged by the scientific community and the government to be unacceptable for use in public school science classes because they violate the constitutional principle of the separation of church and state.

At times both Christians and Naturalists have declared that they were fighting a *war* over man's evolution and set out to destroy the other side, and occasionally, each has acted that way. These actions have been motivated by and will continue as long as the false and unrewarding goal of defeating the other side and winning the *war* is pursued. Who believes scientists are going to give up their pursuit of the *truths* of godless natural science or that Christians are going to give up their belief in the *truths* to be found from their God's supernatural world? God's authority in his supernatural world and the value of religion for believers is not questioned; it is only the reach of Christians outside of God's supernatural authority and into the natural world of Nature that causes conflicts that are subjected to questioning.

Having Christian authors attempting to use godless evolution in their arguments for God raises a related question: Can a Christian who thanks God be a Darwinian who thanks Nature? This, of course, is not a new question, and it has been addressed by many in the past.

For the book, I have selected the discussion mode of independence, for there is less utility in the other modes (conflict, dialogue, and integration). An overview of why these other modes are not used is given in Appendix A—Modes of Interactions.

Who to Thank?

For over two thousand years, Christians have proclaimed that their God has revealed through his scriptures the creation of the universe and man. However, over the last few hundred years, scientific and philosophical advances have established that godless Nature is the creator of the universe and man in the natural world. For Christians, such advances in the natural world do not detract from their belief in God's creations in the supernatural world of their religion. If we limit the supernatural gods to only the Christian God for discussion, there are then two candidates whom we can thank for man's creation and evolution:

God (Christian)—A supernatural agent outside of Nature's universe, defined by the Christian scriptures, communities, and believers.

Nature—A natural process; the sum of all forces, matter and energy in the universe that is manifested by its tools: physics, chemistry, and biology. It is defined by the scientific community.

GOD

With the Christian God selected as the one to thank for evolution, the answer depends on the Christian's interpretation of his God's actions in man's creation noted in the biblical scriptures. Fundamentalist Christians take a literal reading of Genesis and, accordingly, believe in God's instant (a few days) creation of the universe and man. Many other Christians instead interpret the Genesis creation story metaphorically.

Further, the concept of evolution as a process is completely absent in sacred Christian literature. Many biblical scholars argue that the Bible was written to address man's spiritual salvation and guidance and not natural science. To them, quibbling over the words "creation" or "evolution" from a scientific standpoint is missing the point of the scriptures. Accordingly, viewing Genesis with scientific arguments should not be done in the first place. Liberal Christians, although referencing the same God as Fundamentalists, ignore the Genesis time frame and use billions of years and go beyond biblical creation and propose a God-centric version of man's evolution. This requires modifying Darwin's Theory and inserting God into it to support their argument that God works through Nature's Darwinian evolution. Dowd and a majority of the Christian authors (five) have taken this approach and have God involved in evolution as the designer and creator. With these changes to Darwin's Theory they then proceed to *Thank God for Evolution*, as Dowd's title proclaims.

This is all well and good when the invented God-centric theotheory remains in the discussions within the supernatural world of their God—that is, in church sermons and religious forums and literature. However, when Christians move their God-centric theotheories of evolution from the supernatural world of religion to the natural world of secular public education, then conflicts occur.

These conflicts have been observed nationwide for years from Christian attempts to inject various God-centric theotheories, including the widely promoted Intelligent Design theotheory, into public schools. All of these attempts have failed the tests of both natural science and public courts. The scientific community says theotheories fail for not being natural science, and the courts say they fail for violating the separation of church and state.

Since Christian denominations are not inclined to change the scripture's words on creation to include evolution, the discussions on evolutionary theotheories have been left to individual Christians, such as Dowd and the other Christian authors, who are free to invent God-centric theotheories of evolution and seek approval of their invention from their church and other believers. Five of the Christian books reviewed are attempts to get approval of their theotheories from the public even while knowing it will not be forthcoming from the scientific community.

The sixth author is in a third category, Progressive Christians, who accept Christian scholars' advice and keep Genesis free from scientific discussion and consider it a religious narrative. They accomplish this by holding to the belief that in the Christian scriptures, the creation of man is in God's supernatural world, and, in parallel, accepting that there is a separate world-view, the natural world, in which science theories, such as Darwin's Theory of evolution, reside. The supernatural and the natural worlds of this dual view are held to be separate and independent. This is a personal belief, not one supported by a specific church, but it is not questioned by science, for supernatural beliefs are separate from and not addressed by natural science.

The detailed description of the Christian God's acceptance of evolution is in the eyes of the believer whose faith may range over a wide spectrum. Here we will view the Christian God's position through the eyes of six authors.

NATURE

If we select Nature as the agent of evolution, the philosopher Timothy Williamson[7] provides a working insight:

If we say it is the world of matter, or the world of atoms, we are left behind by modern physics, which characterizes the world in far more abstract terms. Anyway, the best current scientific theories will probably be superseded by future scientific developments. We might therefore define the natural world as whatever the scientific method eventually discovers. Thus naturalism becomes the belief that there is only whatever the scientific method eventually discovers, and (not surprisingly) the best way to find out about it is by the scientific method.

For Naturalists, the answer to the question, *who to thank for evolution* is Nature, for they consider evolution to be a natural process occurring in the natural world—or as Williamson said, the world of matter. Nature is treated as an agent comparable to the role given to God.

Most evolutionary biologists embrace the results of the Darwinian Revolution and have concluded that Nature is responsible for the creation and evolution of all living organisms, including man. Darwin's Theory is supported by an overwhelming amount of scientific data describing the 3.5 billion year period of biological evolution on Earth. Naturalists, or as the methodological philosopher Richard Richards notes, science neither rules out God nor assumes God. Further, science does not address the supernatural, for Nature works independently in the natural world. But in thanking Nature, Richards reminds us to be careful and not to personify Nature, for "we aren't metaphorically thanking evolution for some person or a thing as Nature, just the concept of Nature acting as the process." Evolution by Darwinian natural selection is the outcome of natural processes.

Whose Authority?

Who has the authority over explaining man's evolution? From his beginnings, man's innate inquisitiveness has led him to understand how things work by searching broadly for knowledge on Nature and God. Over time, the new knowledge helped his survival. In his early days, man learned how to fabricate better stone hand tools and become a more efficient hunter, how to employ religious rituals to form more cohesive groups, and how to select tribal leaders to efficiently organize the group to defend their turf. Each in its own way increased man's chance for survival.

As a small animal on a small planet, man has been able to learn over time a little about Nature's resourcefulness, the diversity of life on our planet, the vastness of the universe (the very large), and the strangeness of the quantum domain (the very small). Man has achieved this knowledge of Nature through the use of the scientific method to gather data on which he formulates scientific theories for comprehension. This process, scientific methodology, with its built-in checks and balances, has provided man with the ability to ask probing questions about Nature and man's participation in it. Nature's scientific authority has been established through the many successes of the scientific process.

In parallel, man explored the mysterious, supernatural, spiritual world of the gods, which appeared through personal revelations and spiritual experiences. The vast array of gods encountered ranged from fire gods, sun gods, polytheistic gods representing wine, women, and song who cavorted with earthlings and on, to a monotheistic, omnipotent, unknowable God from beyond Nature. The actions of the gods did not always conform to scientific requirements, but their addition to tribes added cohesiveness and over time aided man's survival. Those gods and religions that did so moved on to compete another day. Many religions have emerged, and some have been discarded.

During the last hundred thousand years man's spiritual life and his survival has been mostly guided by Pagan gods, for it has only been in the last three thousand years that the mono-theistic Judeo-Christian God has become sensible. In these few thousand years, the Christian God's authority has been accepted by believers, first in the West, and subsequently expanded to other religions throughout the world. Theological soundness has been determined by the temporal power of those religious organizations surviving.

The Greeks, through their musings on the gods, gave us an insight into their philosophical concepts of god, Nature, and the State: the Mythos, Logos, and Demos. Logos (reason or science) is man's pragmatic mode of thought that enables one to correspond with the natural world and to function effectively without interactions with Mythos. Mythos (mystery or religion) addresses human conditions, man's interactions with fellow man in this world, the search for mortality, and the ultimate meaning of life. Mythos employs narratives that make no pretensions to historical or scientific accuracy, but they are narratives that are instructional and shed light on humanity, the human condition, morality, and values. Demos is the group or state within which man pursues Logos and Mythos. To the Greeks both[8] Mythos and Logos are essential to man's search for answers to life, and neither is superior to the other. Demos is added for answers from Logos and Mythos are given in context of an individual belonging to a group or state.

The authorities of each of these three disciplines are not in conflict when each operates within its own domain, but they do conflict when they overlap.

THEOLOGICALLY SOUND

Obtaining Christian authority or theological soundness on an issue is a difficult, if not an impossible task, for Christians have not been able to agree among themselves on many issues of their religion over its two-thousand year history. This includes even basic religious issues such as: What did God say? What did Jesus say? Is there a Trinity? Was there a global flood? How did God create man? Which of the several versions in the scriptures should they use for God's commandments? What roles can women have in religion? Even the creation of Adam and Eve is given in two different versions in the scriptures, so which one should they believe?

Differences in theological issues have been apparent from the beginning of Christianity. It took over seven hundred years to reach an approved orthodoxy for key precepts of the church with the Council of Nicaea in 732 AD. The Gospels of Thomas, a gospel contemporary to those of the currently accepted gospels, was declared to be unsound theology by church fiat and, accordingly, not included in the approved Bible. The Protestant Reformation in 1517 further divided Christian views. These differences provide bases for groups of Christians to split into the many different Christian denominations, each holding different views on the scriptures.

Further Christian "theological soundness" has varied over time on many subjects. The American Civil War was fought with

Christians on each side arguing that their country's position on slavery was "theologically sound," with each side quoting biblical references for their proof. Before the war, individuals such as William Lloyd Garrison were uncompromising and outspoken critics of slavery and hypocrisy. They were especially critical of church complicity with slavery, not only in the South but also in the North. However, many Northern Christian denominations refused to condemn slavery or sever ties with Southern slave-holding congregations. Near the end of the civil war, the United States declared slavery to be constitutionally unacceptable, although the Christian Bible remained unchanged and continued to have slavery being "theologically sound."

Who is the authority within Christianity to decide which positions in the Bible are "theologically sound" and which are not? For Catholics it is their pope, but for other Christian denominations, it is their leader or the will of the congregation. But there are hundreds of denominations, so there are many "theologically sound" positions to choose from. There is no person or organization by which an overall Christian authority is determined.

The variability of beliefs within Christianity stems from the degree of personal involvement, enlightenment, understanding, and acceptance of the faith among the members. A theologically sound position comes from the individual's precepts of his or her particular point of view.

Denominational variation continues even today as new denominations appear and declare that they are also Christian, with each bringing yet another theological position. There is no process within Christianity and its many denominations, as there is in science, to declare that one view is sound and the authority and another not.

SCIENTIFICALLY SOUND

Scientific revolutions have begun to establish Nature's authority in the natural world by peeling off many church dogmas declaring God's authority over Nature. The Copernican Revolution removed God as the authority describing the movements of planets and the position of the Earth in the heavens. Once the theory was verified by experiments, Nature was given the authority, and the helio-centric theory became the "scientifically sound" theory to describe planetary movements in the natural world. The old "theologically sound" Earth-centric theory was dropped.

Since then many other religious dogmas have been removed by newly discovered scientific theories. For example, two hundred years after Copernicus, Benjamin Franklin conducted tests on lightning bolts in the heavens during thunderstorms by famously flying kites with metal keys and found that the electrical charges he collected on the keys attached to the kites were the same as the electrical charges he could make by creating static electricity on the keys in his home. His home-made electric charges were the same as those in the heavens and scientifically repeatable—no God needed. With that data, scientific theories replaced "theologically sound" theotheories describing God's creation of lightning. But even today some Christians hold the authority of God over other weather phenomena such as rain, hurricanes, and tornadoes, all of which have also been explained as natural events.

If we ask the question *Who has the authority over the weather, God or Nature?* many Christians today would say God, for scientists have difficulty persuading many believers to accept the case for Nature. Christians may hold that the weather and any other physical phenomena are caused by God, for it is their right to believe in God's supernatural narrative. The Bible refers to supernatural lightning fourteen times. Naturalists say fine for sermons, but when Christians attempt to inject supernatural beliefs into the natural world, they are violating the rules of the scientific community, and such supernatural arguments should be rejected.

The same is true for Darwin's Theory of biological evolution. The data supporting the evolution of man through the process of natural selection are overwhelming, and the scientific community has long since declared the soundness of the theory. However, having most Christians accept Darwin's Theory has proven to be most difficult, for there are still Christians working (devilishly?) hard to rewrite the godless Darwin's Theory with God-centric theotheories to explain evolution. In fact, explaining the difficulties Christians are having establishing their authority over evolution is covered by the critique of several Christian authors in this book.

As issues get distant from the tests, human interpretations of science tend to creep into the reporting and invalidate the scientific soundness. Without a strong link to tests, scientific soundness disappears.

Resistance by Christians to Darwin's Theory has continued for over 150 years and has resulted in a backlash by scientists and philosophers who argue that Darwin's natural science theory is not subservient to the supernatural Christian God, for they believe that Nature is independent of God. This backlash has resulted in the publication of a number of books by scientists and philosophers, known as the new Atheists, defending science and rebelling against believers forcing God's intrusion into science. New Atheists' books include Richard Dawkins' *The God Delusion*, Christopher Hitchens' *God is not Great*, Daniel Dennett's *Darwin's Dangerous Idea*, Jerry Coyne's *Why Evolution is True*, and Victor Steger's *The New Atheism*. All of these books argue that God has no authority over science or governance in the natural world.

JOINT SOUNDNESS

Looking for an issue to be both theologically and scientifically sound is a most improbable task; yet there are some Christians who say they can do this. Lisa Randall[9], a leading particle physicist, outlines the problem:

> *Religion might well yield valuable psychological benefits. But any religious scientist has to face daily the scientific challenge to his belief. The religious part of your brain cannot act at the same time as the scientific one. They are simply incompatible.*

To illustrate the basic conflict when attempting to join a "sound" scientific and a "sound" theological position, we do not have to stray from the issue at hand, evolution. The fundamental tenets of the creation-evolution of man for God and Nature are antithetical:

> *Theologically sound*—God created the heavens, the Earth, and man and designed man in his image.

> *Scientifically sound*—Nature provides the processes by which the heavens, the Earth, and man were created without a creator and designed without a designer; no gods required.

There is no logical way in the natural world to construct by mixing, marrying, or building bridges a sound common position between these two vastly different positions on the creation, evolution, and design of man. Forcing God to be scientifically sound would remove him from his supernatural narrative describing creation. Forcing Nature to be theologically sound would violate Darwin's Theory. For example, removing chance (theology says God's direction cannot be by chance) would weaken Darwin's Theory for the process of natural selection functions with chance as one of the defining processes.

Another example is found in the God-centric theotheory recently used by a Catholic cardinal to describe the creation of the universe. Cardinal Schonborn gave the Vatican's position on the creation of universe by quoting from the Christian narrative: "In the beginning God created the heavens and the earth" (Genesis 1). If the cardinal's pronouncement was directed to believers, there would be no critique, but he directed his comments to the population at large announcing his imposition of his supernatural views on the wider world.

These are, essentially, the first words of religious instruction for belief in God the Creator and belief that he created the heavens and the earth. These instructions are the foundation on which every other Christian belief rests. To believe in God and, at the same time, not to believe that he is the Creator would mean, as Thomas Aquinas puts it, "to deny utterly that God is." God and Creator are inseparable. Every other Christian conviction depends on this: that Jesus Christ is the Savior, that there is the Holy Spirit, that there is a Church, and that there is eternal life: they all presuppose belief in the Creator.

These "theologically sound" statements are part of the supernatural Christian narrative, essentially a quote from the scriptures, and as the cardinal says, an article of faith. But faith in the supernatural world is not acceptable in the natural world of science, for it conflicts with Nature's theories of the universe on almost every point. In the natural world, Nature is the creator of the universe; there is no common "soundness" ground between these the two worlds. However, if God the creator is understood to be in a supernatural religious narrative, there is no conflict, for it is a religious belief separate from, but acceptable, as all religions are in the Nature's world.

On the personal level, the same problem occurs—the inability to arrive at a "theologically sound" Christian positions within Nature's world. Take the subject of homosexuals. who are discussed in the scriptures and placed in the same category as fornicators, idolaters, and adulterers:

> *Do you not know that the unrighteous will not inherit the kingdom of God? Do not be deceived. Neither fornicators, nor idolaters, nor adulterers, nor homosexuals, nor sodomites.*—1 Corinthians 6:8-10

> *Thou shalt not lie with mankind, as with womankind: it is abomination.*

> *For whosoever shall commit any of these abominations, even the souls that commit them shall be cut off from among their people.*— Leviticus 18:22-29

Social scientists today argue that homosexuals do not appear to have more drunkards or to be more wicked than the population at large. US law[10] does not allow for homosexuals to be treated any differently from other citizens. However, some Christian denominations have discriminated against them even though these actions are against existing laws and our scientific understanding is that sexual orientation is not a choice, but rather a complex interplay of biological and neurological factors.

Although Christian scriptures argue that homosexuals be cut off from their people, this has not been adopted by US law. Clearly Christian moral values when considering homosexuals are immoral today under our secular constitution. Some churches have welcomed homosexuals into their congregations and priesthood, and have even elevated them to bishops of their church. Other churches have taken the opposite positions by banning homosexuals from priesthood and any other position of power. This issue has dramatically split the Anglican Church, which is currently being torn apart by the two opposing views, each arguing that it is the "theologically sound" denomination. What is the theologically sound position on homosexuals within Christianity today when even denominations cannot agree?

Much Christian dogma can be explained in light of history. In early tribal cultures, homosexuals would have been a small minority (as they are today) and would be seen as different, outside of the norms of the majority and not contributing to the biological growth of the tribe. Seen in this context, it could be argued that homosexuality was immoral, as it could jeopardize the survivability of the tribe. The Bible was written reflecting the tribal views of that time and place, and, accordingly, homosexuals were "abominations," that is, they were different people who had to be rejected for not conforming to common expectations of the tribe to produce children needed for the group's survival. But those conditions—the necessity of high birth rates—are not the conditions of today.

This raises the question of who defines a theologically sound position in biology. An example of a "theologically sound" position in conflict with science was the early mishandling by many Christian denominations of the HIV disease. Some churches have a long-standing dogma forbidding any form of contraception, with the exception of abstinence. Along came the latest sexually (primarily) transmitted disease, HIV, and public health officials immediately recommended the use of condoms to prevent the disease from spreading. The Catholic Church made the dogmatic judgment that it was more important to defend their dogma against condoms than to save lives by their use. This opposition has continued in the face of real world data showing that condoms could have prevented a large number of infections and deaths, while abstinence only has been of little or no help. However, some churches and many believers have accepted the science and supported the use of condoms.

There are, however, times when theological soundness can appear to be the same as scientific soundness. Example: the moral prohibition of stealing is both a biblical moral given by God and a cultural moral that aided evolution. But this sameness, Naturalists would argue, is because biblical morals were selected from the societal morals that had evolved before biblical times.

A strength of Christianity has been that it is a "big tent" organization with many different supporting posts—that is, different points of view based on biblical interpretations of Christian theology by the hundreds of denominations. With wide diversity and no overall selection process, there is no one "theologically sound" position for all Christians.

Thus, attempting to promote any theotheory that is theologically sound across all denominations and all Christians is doomed to failure before it gets out of the first church door. Incorporation of miracles dooms a theotheory by scientists before it gets in their labs. These conflicts in togetherness highlight the need for the recognition of the independence of religion and science. Religions should have the independence to debate theologically sound positions without critical review from science.

It is a pity that Jefferson, Madison, and other Founding Fathers, while constructing the constitutional wall separating the state from the church, did not also build a wall separating science from the church. But our Founding Fathers were engaged in their constitutional activities almost a hundred years before the Darwinian revolution and can be forgiven. However, the state has ruled that in public school classrooms, biblical science statements are a violation of the existing wall separating religion and state, whereas secular natural science is not. Court rulings based on this constitutional wall of separation of church and state have succeeded in most cases to keep biblical science out of public organizations and natural science in.

Which Theory?

We can postulate four contending hypotheses for answering the question of *who to thank for evolution—God or Nature?* Each hypothesis accepts, modifies, or rejects Darwin's Theory of evolution. Naturalists accept the theory directly: Darwin's Theory is godless. For Christians this poses a problem for they must decide how to accept a godless theory that conflicts with their supernatural God narrative. There is a spectrum of Christian beliefs concerning the conflicts with evolution. For discussion we will select three Christian groups to represent their belief spectrum: Fundamentalist, Liberal, and Progressive. Their approaches are compared with those of non-theists, the Naturalists.

God Hypothesis—God has authority and power over all worlds, including the world of Nature. God's biblical Genesis describes the creation of the universe and man for all worlds, natural and supernatural. Darwin's Theory is rejected.

> This is the preferred hypothesis for Fundamentalists who adhere to a literal reading of Genesis.
>
> God-over-Nature Hypothesis—God has authority and power over God's supernatural world and the authority to

THANK EVOLUTION FOR GOD

act through Nature's process, including those described by Darwin's Theory.

This is the preferred hypothesis for Liberal Christians who replace Darwin's Theory with theotheories that modify Darwin's Theory to be God-centric.

Nature-plus-God Hypothesis—Nature has authority over the natural world, including the creation and evolution of the universe and man. God's narrative is accepted for the supernatural world and kept separate from the theories of Nature, including Darwin's Theory.

This is the preferred hypothesis for Progressive Christians who accept Nature's world and separately believe in a supernatural narrative with God.

Nature Hypothesis—Nature has authority over the natural world, including the creation and evolution of the universe and man, and creations by man, such as supernatural narratives with a God. The supernatural narratives are kept separate from the theories of Nature, including Darwin's Theory.

This is the preferred hypothesis for Naturalists who fully accept Darwin's Theory and have no need to address the supernatural, but they do accept that one may also believe in the Christian narrative.

Selection of the God Hypothesis, also known as Creationism, by Fundamentalist Christians, is a satisfactory answer for a minority of Christians *who thank God for creation*. For Liberal Christians, this is inadequate and out-of-date with science, including Darwin's Theory, so they extend God's authority to include Nature. By mixing God and Nature they modify science theories, such as Darwin's to be God-centric. Such modifications by theotheories put God in charge of evolution and extend the evolutionary time period to billions of years. These adjustments are their basis for declaring *Thank God for Evolution*. Since any invented supernatural theotheory with God involved will be different from Darwin's Theory, these Christians knowingly accept conflict with and rejection by the scientific community on evolution.

And indeed the mixing of religion and science has been used in past theotheories, such as Scientific Creationism and Intelligent Design, which have failed in public court houses and science classrooms. To circumvent these failures, additional God-over-Nature theotheories have been proposed, such as Theistic Evolution by Francis Collins and Sacred Innovative Design by Dowd.

MYTHS OF TOGETHERNESS

Knowing that Darwin's Theory is godless, many Liberal Christians try to find arguments that join their God with godless natural science to remove conflicts. They try to achieve togetherness by bringing religion and science under God's authority through the development of theotheories which declare (1) it is acceptable to "marry" or mix science with religion, (2) it is acceptable to build "bridges" between science and religion, and (3) it is acceptable to modify scientific theories, such as Darwin's Theory, as directed by God.

These myths fail the tests of science and are rejected by the scientific community. They misinform Christians what science is all about.

A summary of several of these myths of togetherness or fear is outlined below.

CHAOS

There is much Christian literature projecting the potential chaos within the world's population that would follow if one does not believe that God is the creator of the universe and man. Further, as the chaos argument goes, the removal of God as the cause would mean a victory for Atheists over Christians. And indeed militants on both sides have highlighted fears of what happens if the other side wins the *war*. On the Christian side, there is fear that God will be pronounced dead or that the ethical and moral principles derived from religion will be rejected. Ken Miller, a Catholic scientist, summarized these fears in his book *Finding Darwin's God*. With the advance of science, specifically evolution by natural science, he worries that:

> . . . the real risk is that evolution tells people that God is dead. And if people were to believe that, they might indeed behave as if all is permitted and social chaos would result.

Or that morals would be lost:

> . . . but if ordinary people were to discover that the ethical and moral principles derived from religion were nothing more than a convenient social fiction, all hell might break loose. They might behave as if anything was permitted, and society would come apart in a flash.

Although some militant Atheists have used comparable arguments, science cannot say whether God is dead or not, for it simply does not address supernatural Gods. Whether God is dead or alive can only be answered by Christians, not non-believers, for it is a belief which does not require proof. There is little risk that believers will be swayed with statements from non-believers addressing their supernatural God.

The same argument holds for ethical and moral principles. Believers are free to embrace Christian morals or those of any religion regardless of what non-believers say.

COMPATIBILITY

Many well-meaning Christians wish their religion and science to be compatible. The wish leads to inventing reasons why this is so. One example comes from a well-known scientist, Stephen Jay Gould, who invented the concept of non-overlapping magisterial—a concept that holds that science and religion each has "a legitimate magisterium or domain of teaching authority," and that these two domains do not overlap. No overlap, no conflict. Although simple in concept, it is found wanting when put to a test of logic. For example, the question of *Who created the universe?* is obviously a question that can be addressed in the magisterium of each, but will produce vastly different answers, one by a supernatural God and the other by Nature. Obviously the questions and their answers can overlap and if so they can be in conflict. However, if Gould had added that the religious information is supernatural and Nature's information is natural, they can be said to not conflict, for science does not address the supernatural.

MARRIAGE

Theotheories that "marry" religion and science are being put forward by Christians as replacements to Darwin's Theory by mixing religiously acceptable actions by God, who can do anything with the tenets of Darwin's Theory. Having God direct evolution violates Darwin's Theory and the independent relationship of religion with science. The authority for science theories is the scientific community, and they do not allow religious modifications of science theories. Some may wish to call it a "marriage," but in reality it is an unacceptable insertion of the supernatural into natural science theories in order to make them God-centric. Such "marriages" provide only a veneer of science to religion and bad theotheories to science. Both are distractions to be avoided.

The assumption that God can be mixed with Nature as Christians see fit may be convenient and acceptable for a Christian's personal belief and Sunday sermons, but it is not an acceptable position for use in natural science classes of public schools. The independence of Nature and God needs to be honored.

BRIDGE BUILDING

At times Christians and philosophers have attempted to construct metaphorical "bridges" over the scientific gap between Christian religious views of evolution on one side and natural science's position given by Darwin's Theory on the other. Unfortunately, such "bridges" are not constructible because the scientific community rejects any bridge foundation with supernatural elements of God (bad science) and the religious community rejects Nature's godless elements (bad religion). Oops, there goes the bridge into the canyon.

Marriages, bridges, and harmony between science, religion, and the state are not to be expected, nor desired, for we are far better off being served by independent views: those of the natural world (science and state) and that of the supernatural world (religion). We need the wisdom from each, and those in Heaven and Earth know that man needs all of the help he can get to solve our real world problems.

SEPARATION

Many scientists have recognized that religion and science are fundamentally different and should not be mixed together. Science organizations, such as the National Academy of Sciences in their publication *Science and Creationism*, have noted this:

> *Scientists, like many others, are touched with awe at the order and complexity of nature. Indeed, many scientists are deeply religious. But science and religion occupy two separate realms of human experience. Demanding that they be combined detracts from the glory of each.*

Arguments for togetherness only defer discussions on workable solutions to the conflicts into the future. These theories should be known as "kick-the-can-down-the-road" theotheories, for they try to please some Christians today, but only by pushing real solutions to a later time.

The necessity for the separation of the two realms is explored in detail through the analysis of the failures of five Christian authors using mixtures of religion and science together to remove the conflicts between religion and Darwin's Theory of natural selection.

What Information?

We live in a world, the natural world, where gossip, opinions, facts, theories, religious revelations and beliefs, state laws, and religious theotheories are swirling about us coming from many different sources: books, newspapers, talking heads on TV, Internet blogs, universities, churches, and the government. What to make of it all? For answering our question on evolution, two categories of information are important: that in the natural world and that in the supernatural world.

To provide a starting point on how to separate the facts from the opinions, the natural from the supernatural, and natural theories from religious theotheories, we start with a definition from the National Academy of Sciences on religion and science:

> *Science and religion are based on different aspects of human experience. In science, explanations must be based on evidence drawn from examining the natural world. Scientifically based observations or experiments that conflict with an explanation eventually must lead to modification or even abandonment of that explanation. Religious faith, in contrast, does not depend only on empirical evidence, is not necessarily modified in the face of conflicting evidence, and typically involves supernatural forces or entities.*

Because they are not a part of Nature, supernatural entities can-
not be investigated by science. In this sense, science and religion are
separate and address aspects of human understanding in different
ways. Attempts to pit science and religion against each other cre-
ate controversy where none needs to exist.

An interesting historical note to the NAS statement is that
one of its founding members, the Harvard scientist, Louis Agassiz
(1807-1873), after first hearing of Darwin's Theory, argued that
Darwin must be wrong, for God would not create the living world
by biological selection, by random variation, and survival of the fit-
test. Further, he argued that one's view of life must not be allowed
to descend into slimy ponds. However, since that time science has
moved on, many experiments have been conducted, and Agassiz has
been proven wrong. Today, even at Harvard, scientists fully embrace
Darwin's Theory and its view that man's ancestors are indeed from
those slimy green ponds. Living organisms have evolved by chancy
DNA mutations and random variations using the process of natural
selection that allowed life to climb out of the slimy green ponds, walk
on the land and soar into the skies.

Unlike Agassiz, scientists now separate religious information
from scientific information and don't try to tell God what he likes or
dislikes. Scientists let the experimental data speak for itself. Beliefs in
God are kept separately in God's supernatural world. Some Christians
still agree with Agassiz's arguments on why Darwin must be wrong
today, even though the advances of biology have given us much data
to argue otherwise.

The NAS statement acknowledges that "science and religion are
separate and address aspects of human understanding in different
ways"—God's way in the supernatural world and Nature's way in the
natural world.

As noted, our discussions have been broadened to include the State as a third independent discipline. Our democracy, the State, is a source of information and laws and has its own process of checks and verification in reference to the Constitution. By law, the constitutional laws must be secular.

Theories about the natural world and theotheories about the supernatural world are quite different. This is understandable, for believers in the supernatural world use God as the source for personal supernatural revelations about God. On the other hand, to understand Nature, which has no God to give guidance or information, man must rely on the scientific method to verify its information and theories. From observations hypotheses are proposed and tested, and for the successful ones, theories are constructed.

Information from God's supernatural world is accessible by believers. Beliefs held by Christians or other believers are not questioned, for they are not addressed by natural science. What are questioned are the extensions of God's authority into Nature's authority over natural science theories. This mixture of supernatural and science information will cause much time to be spent separating the two in order to address whether God or Nature has authority over man's creation-evolution.

God, Nature, and the State are very large subjects, but only selected topics and limited issues will be discussed in this small book. Quotes from expert sources are used in the discussions to emphasize a point of view and support the dialogue. References and supporting data are listed for readers to pursue key points in more depth.

Discussions on Nature are guided by the discipline of the scientific community. Discussing God's acts is another story, for religious information is administered by multiple denominational communities, many of which often differ among themselves. For example, there are major differences between Quaker, Evangelical, and Catholic Christians on many positions, even though they believe in the same God, almost. This is an understandable state, for religious supernatural beliefs are not testable, do not require rigorous classification or verification, and are given authority status by being declared a dogma by a church. The Christian comments concerning their God employed in answering the question are based as close as possible on the descriptions given by the Christian authors, such as whether God uses miracles with a proposed evolution theotheory.

Using generalizations such as the "scientific community," "scientist," "Fundamentalist," or even "Christian" can be hazardous, for they reflect the author's selection of the mean opinion in that subject knowing that most of the subjects being discussed have a spectrum of opinions. However, the use of a specific position is necessary at times to facilitate discussions while knowing that selections of the extreme positions could lead to different conclusions. Hopefully, the statements accurately reflect a majority of scientists. This, of course, does not guarantee that it is the "truth" or even that scientists know what the "truth" is.

Women may wish to scold the author for the use of "man's evolution" rather than "the evolution of humans," but I argue for simplicity.

The use of quotes to support arguments can also be hazardous, for quotes can be taken out of context when supporting paragraphs around the quote are missing. These hazards are recognized, but there is far too much ground to cover without the help of such shortcuts. Further, in quoting the scriptures, there is always the additional hazard of choosing between a metaphorical or a literal meaning. Example: Noah's global flood may be viewed as an important literary metaphor that does not consider the underlying science needed to explain a flood of forty days over the entire earth with a water depth of thirty thousand feet.

Scientists, including Darwin himself, prefer describing his theory as the theory of natural selection rather than evolution, but since most of the public and most references refer to the theory of evolution, these two are used interchangeably throughout the book.

Book Path

There are two world-views, Nature's material world and God's supernatural world, being studied for answers to the central question posed: *Who to thank for evolution—God or Nature?* The case for Nature, or *thank evolution for God*, is summarized in the chapter Nature's Way. The impact of the State on religious and scientific views is given in the chapter The State's Way.

The case for *thank God for evolution* requires a broad scope, for with Christianity there is a spectrum with many different versions of Christianity. For simplicity I represent this spectrum of beliefs by three groups: Fundamentalists, Liberals, and Progressives. The focus is on Liberal Christians with the views of five different Christian authors used to thank their God and define the relationship with evolution. A sixth author presents the Progressive view. Each of the six books is summarized in light of Darwin's Theory separately in order to highlight each author's explanations of God's authority with Nature's evolution. Answers to the question *who to thank for evolution—God or Nature?* are derived for each author in light of the acceptance of the science of Darwin's Theory. Discussing each author separately presents some redundancy, but allows for a more in-depth understanding of each author's views on the interaction of supernatural religious concepts with natural science and the arguments on how to overcome the conflicts created. Finally, a solution to the conflicts between religion, science, and the state is summarized in the chapter Partnership.

OTHER BOOKS

Other books were considered, but these six captured the essence of the question. An example of a book not included is Kenneth Miller's *Finding Darwin's God: A Scientific Search for Common Ground between God and Evolution*. Miller's excellent technical description of the biology of Darwinian evolution is covered by one of the accepted authors, Ayala, and no new evidence is presented to support his Christian views of God's creation-evolution of man. Amiel Rossow has given an excellent, in-depth review of his book in *Yin and Yang of Kenneth Miller: How Professor Miller finds Darwin's God*. Rossow notes that although Miller is a scientist, he fails to find any scientific evidence for God:

> *I don't believe skeptics will be swayed by Miller's pro-faith arguments. This is not because his arguments {for a scientific basis} are doubtful or weak, but simply because there are no arguments at all, just assertions not supported by evidence but repeated time and time again with a boring persistence.*

Further, Rossow notes that:

> *When he writes about evolution {science}, he strictly adheres to well-substantiated evidence. When writing about his faith, he does not offer any evidence, appealing instead to assumptions lacking parsimonious content.*

Miller is one of several Christian scientists trying to provide convincing scientific arguments that Darwin's Theory supports his faith rather than be in contradiction, but one is left with the simple observation that contradictions are removed only by separation.

Nature's Way

Nature is us and all that is around us—the stars in the cosmos, gorgeous vistas on Earth, myriads of incredible bacteria, insects, and animals, including humans—two million living species in profuse diversity. This is only a part of Nature, for we don't readily see many things vital to our lives, such as the colonies of bacteria living within our bodies and the soil, which, if they were not there toiling away, we could not live. Ancient fossils in museums remind us of life's long and varied history, going back about 3.4 billion years. Molecular biologists through DNA research are comparing man's DNA, the instructions for life, with some of our earlier ancestors' DNA that has been found (luckily for the case of Neanderthals) in a few fossils. From this DNA and from the DNA of living species, a step-by-step account of the evolution of many of man's ancestors is coming into focus in digital clearness.

But much remains to be learned, with many questions remaining: In biology, how was the first replicating cell created, and is there life on other planets? In physics, how did matter win out over anti-matter after the Big Bang, and what is future of the universe in light of the acceleration of its expansion?

Observatories with telescopes are giving us detailed pictures of planets orbiting around our Sun and even tantalizing peeks of planets orbiting around distant suns and views of billions of galaxies billions of light-years away. Physics labs have particle accelerators smashing nuclear particles together, which give us brief peeks at the quantum world's weird building blocks of matter, including that which existed shortly after the Big Bang, such as, the newly discovered Higgs particle that existed a billionth of a second afterwards. Slowly we are learning about how Nature's tools created and designed the vast universe and the living organisms on this Earth.

We can picture scientists in their laboratories working away to develop predictive pictures of how the natural world works. Its value to individuals and the community depends on acceptance by the State and religions. Down the street from the science laboratories, there are state buildings in which we can listen to officials debate secular state laws that define civic morals governing our daily lives and the acceptance of natural theories. A block farther down the street, there is a church with a minister giving a sermon on man's creation and morals given by a supernatural God described in a religious narrative two thousand years old. The minister is preaching that Christianity offers a common thread for the vistas, personal moral guidance and salvation—all miraculously given by his God.

Scientific revolutions have been able to give scientists the information to see a different common thread tying these widely varying scenes together—they are all products of Nature—which we have learned about from intellectual revolutions. Four hundred years ago the Copernican Revolution gave us views of the workings of the cosmos and over time further observations and experiments have given us data on the birth of the universe from the Big Bang. The Darwinian Revolution has advanced evolutionary biology and given man a deeper understanding of the common thread of DNA among all life and descriptions of the diversity of life. The Neuronian Revolution is beginning to describe the evolution of man's brain with the mental capacity to construct tools, moral codes, social governing systems, scientific theories, and supernatural God narratives. Armed with this knowledge, we should be able to address the question, *who to thank for evolution—God or Nature?*

So it is only to be expected that with different models of the origins of man and the universe, one by science from Nature's world and one by belief from God's supernatural narratives, there are conflicts. This section presents the case for Nature. Separately the case for God is presented later in the chapter God's Way, and the impact of the State on science and religion is outlined in the chapter The State's Way.

BEGINNINGS

Our understanding of Nature comes from observing, finding patterns, proposing hypotheses to explain, and then conducting experiments to prove or disprove the hypotheses. This is known as the scientific method. Hypotheses that fail this test are rejected; while those that pass are used as a base of knowledge to model Nature until the next set of observations indicate a difference, thus starting the learning cycle over again. One never knows how long our present-day theories will stand. In science studies of Nature's world, the supernatural and miracles are not allowed, for they are but a tacit acknowledgment of "I don't know"; specifically, "I don't know how to describe the event with Nature's tools available, so I will invent a description or theo-theory with help from God." For scientists it is acceptable to say, "I don't know" and then live with the uncertainty and absence of knowledge while working (like the devil?) to find an answer. An old cartoon by Sidney Harris describes this stage:

Two scientists are standing before a blackboard on which one scientist had written a long equation, but with one missing term which has been replaced by the words "Then a miracle occurs." The second scientist notes: "I think you should be more explicit here."

In explaining unknowns, a religious believer will accept a miracle and an unverified supernatural explanation attributed to his God, while a scientist will go back to the lab and gets more data in order to be more explicit in his understanding. Scientists know that any answer given today includes uncertainty, but it's the best that can be done now, so let's get on with it.

Some of the uncertainty we live with today from observations indicate that we:

. . . live in a universe in which a mysterious force known as dark energy makes up about 70 percent of the total cosmic amount of everything. A mysterious substance known as dark matter makes up about 25 percent. And ordinary matter—the stuff of the periodic table, including interesting assemblies of matter like galaxies, stars, planets and people—is a paltry 5 percent.

This is the macro-universe where Einstein's theory of relativity works. But there is also a micro-universe, the quantum universe, where relativity does not work and quantum and string theories present our best understanding. But the theories in the large and the small realms do not work in unison so well with our present understanding of Nature. Kenneth Ford in his book *101 Quantum Questions* gives a description of the quantum world in the eyes of John Wheeler, a major contributor to the field:

The way {the quantum world} is—probability, uncertainty, waves and particles, superposition and entanglement, and all.

Neither he {Bohr} nor any other researcher needed to know whether anything deeper lay beneath the theory.

With any theory there is no way for scientists to know if there is another answer out there stranger than the ones we have, for scientists have no God to point the way or give divine guidance to say you have found the answer or even indicate how close we are to an answer. Scientists remove one barrier at a time blocking man's vision and by doing so allow him to see a little further ahead today than he did yesterday.

At one time Christians viewed rainbows, lightning, and comets in the sky as miracles or omens their God sent to inflict pain or pleasure on man. But over the years, each has turned out to be scientifically explainable as a natural event with Nature's tools. Once Isaac Newton performed experiments with light beams and prisms, he was able to literally see the rainbow's color spectrum (a prism splits white light into its different frequencies, giving different colors) and with this data he was able to offer a physics theory for the rainbow. Newton moved our understanding of the rainbow from being a supernatural event created by God to a natural event created by Nature. Now we know that the rainbow can be explained by physics as a natural event, a repeatable experiment and a beautiful sight in Nature to write poems about. We also know that however beautiful it is with its multiple colors and arching shape, it is designed without a designer and gives an image visible in the natural world by obeying Nature's physical laws—no supernatural intelligent or innovative designer required.

Newton was instrumental in establishing the Royal Academy in 1660 in England, a sign of scientific enlightenment, and this only about thirty years after Galileo's trial in Italy for heresy for supporting another natural science theory. The Society's motto, *Nullius in verba*, roughly meaning "Take nobody's word," is fitting for the scientific approach.

Our task is to separate the facts that describe the natural world from supernatural beliefs and theological thinking, or as Charlotte Perkins Gilman captured in her poem:

> *Once we thought the earth was flat—*
>
> *What of that? It was just as globos then Under believing men*
>
> *As our later folks have found it, By success in running round it;*
>
> *What we think may guide our acts, But it does not alter facts*

Peeling off old religious dogmas built on God's supernatural explanations of Nature's world, like those once attached to the rainbow, has been a major task for science for over a thousand years. Removing God's explanation of man's creation and replacing it with natural selection has been a comparable task for the last 150 years since Darwin, but with the added emotional intensity engendered by questioning God's special relationship with man. But the peeling away dogma must be done to understand Nature and the science of natural selection.

The scientific community acts as an overseer to the scientific process, and our joint understanding of Nature serves to insure that we don't fool ourselves by including non-testable theories, such as those with supernatural input (gods, angels, devils, talking serpents, miracles, etc.). Perhaps Richard Feynman's view of science sets the stage:

> *Science is a way of not trying to fool yourself. The first principle is that you must not fool yourself, and you are the easiest to fool.*

It is scientific experiments on which we base our theories. As the theoretical physicist Lisa Randall notes:

Our hypotheses {and theories} are initially rooted in theoretical consistency and elegance, but, as well as we see ultimately it is the experiment—not rigid belief—that determines what is correct.

Nature's way is verified by scientific experiments, and it is from these experiments that we ensure that we are not fooling ourselves.

Religious believers see the same vistas and animals and experience comparable feelings of beauty, plus a connection to their God. They argue that Naturalists without belief do not see God's work. This is true, but this leaves Christians with the difficult task of proving to others, non-believers particularly, that they have not fooled themselves and have separated science from belief, Nature from God. They can see both, but questioning the validity of what is seen is vastly different.

Our understanding of Nature is:

Nature is the world around us that we observe from inert matter and living organisms, from the very large to the very small. Following its own laws (physics, chemistry and biology) Nature has created the universe and living organisms without a creator and designs without a designer. One of the living organisms, man, has evolved a brain with an inventiveness that allows man to explore the mysteries of the universe with science and to create supernatural God narratives.

Nature does not speak to us directly or give us a narrative to follow; man has to acquire any understanding or narratives by himself. Environmentalist Rob Watson outlined an illustrative way of thinking about Nature (edited to add biology):

Mother Nature is just chemistry, physics {and biology}. That's all she is. You cannot sweet-talk her. No, Mother Nature is going to do whatever chemistry, physics and {biology} dictate. Do not mess with Mother Nature.

Learning about Nature requires untangling the history of religions and states, each of which has at times helped and at other times retarded the advance of science and our understanding of Nature.

The arc of the natural history of man's biological evolution starts over three billion years ago with a common ancestor of life, the first life form, and grows by the process of natural selection through many different species: from single-cell bacteria, fish, shark, lizard, mouse, ape, and down to man. Man's branch of the tree is bushy with several parallel branches existing over the last eight million years, each species a contender vying to survive. One made the transition from the ape to chimp to man.

The last few steps beyond our ape ancestors and into the hominini family branch were monumental in increasing the animals brain size and mental capabilities—the hallmarks of modern man. Expanded tool making and social capabilities allowed our ancestors to increase the size of their tribe and communicate with other tribes, thereby expanding the range for social contacts and information gathering for tool making, all of which increased his mental capabilities.

About one hundred thousand years ago, modern man (Homo sapiens) emerged and came out of Africa. Man now had the capability to not only survive but to explore the mysterious world around him: why the sun traveled across the sky and why the rivers flooded each year. Shamans gave supernatural myths with gods.

By about ten thousand years ago, urban centers appeared and later expanded into city-states. States with multiple cities were founded, and the power of their religions and gods grew and fell with the fortunes of the state. Minoans, Greeks, and the mighty Roman Empire appeared, and Western civilization was off and running, with science advancing. After Emperor Constantine established a Christian theocracy in 325 AD in the Roman Empire science also came under its control, and for a thousand years, religious dogmas hindered the advance of science. Scientists in Muslim countries after 800 AD stepped forward and made significant contributions to science.

In Europe after 1400, science began to emerge from the shadow of the church with the assimilation of the work of Muslim science and the emergence of new scientific theories. These theories challenged the standing church dogmas over science. Many of the scientists were in the church and proceeded with due caution, for it was only to be expected that new scientific advances would cause conflicts with the existing Christian dogma and in some cases put the life of the scientist in danger.

The spark that set off the first of the major scientific revolutions was ignited in the mid-1500s by a Christian friar in Poland, Nicolas Copernicus, who proposed a natural theory for the planetary system with the Sun at the center of the universe and not the Earth, a theory that was in conflict with the existing church's geo-centric dogma that had man and the Earth at the center of the universe. At that time there was no independent scientific community to call upon for support, only isolated scientists working independently, many within the church, such as Copernicus. It was a dangerous activity to challenge the church dogmas, but a few did successfully, and new theories began to emerge which would form the knowledge that would allow scientists to peel away layers of encrusted church dogma.

Galileo, a recognized scientist of the time, was an early supporter of Copernicus' theory, and with his newly invented telescope he was able to gather supporting proof for the helio-centric theory. Since Galileo, ever increasingly powerful telescopes have given us glimpses of our universe so that now we have been able to see objects from thirteen billion years ago, a time near the very beginning of the universe. Other scientific tools, such as particle accelerators, have expanded our knowledge of what stuff the universe is made of. Most recently, Laurence Krauss notes that the discovery of the Higgs particle:

> . . . makes even more remarkable the precarious accident that allowed our existence to form from nothing—further proof that the universe of our senses is just the tip of a vast, largely hidden cosmic iceberg.

The advance of Western secular knowledge was supported by a general philosophical awaking in Europe known as the Enlightenment. From this and three major scientific revolutions Naturalists were able to build a foundation of scientific knowledge for challenging church dogma which held that God has the authority over Nature and the State.

In the Enlightenment the works of Bernard Spinoza are examples of the struggles that early philosophers had in Europe with their efforts to break free of the control and dogma of religion. A review of his works is outlined in Steven Nadler's book *A Book Forged in Hell*, in which he quotes Spinoza's explanation of his reasons for questioning the church's controlling position over philosophy:

> I am writing a treatise on my views regarding Scriptures. The reasons that move me to do so are: 1. The prejudices of theologians. For I know that these are the main obstacles that prevent men from giving their minds to philosophy. So I apply myself to exposing such prejudices and removing them from the minds of sensible people. 2. The opinion of me held by the common people, who constantly accuse me of atheism. I am driven to avert this accusation, too, as I far as I can. 3. The freedom to philosophize and to say what we think. This I want to vindicate completely, for here it is in every way suppressed by the excessive authority and egotism of preachers.

Some two hundred years after Copernicus, the Democratic Revolution in America established a government separated from and independent of religion. The democratic government was secular and would not harbor religious dogma that could be imposed on its citizens. Three hundred years after Copernicus, the Darwinian Revolution would establish Nature as the authority over man's evolution, and Darwin's Theory would explain the diversity of life on our planet. Now, a hundred and fifty years later, a Neuronian Revolution being led by neuroscience research supports the hypothesis that man has evolved a moral sense and the mental capabilities to produce scientific theories to explain Nature and to invent supernatural narratives to explain God.

COPERNICAN REVOLUTION

The intellectual darkness that had engulfed science in Europe during the Middle Ages began to slowly lift when in the mid-1500s, a gifted monk of the Catholic Church, who was also a scientist, came forward and challenged the existing dogma of the church's geo-centric theo-theory of the cosmos. As a good Catholic, Copernicus dedicated his work to the pope with the hope that it would enlighten the church by improving the predictions of planetary orbits. His theory would replace that of Ptolemy which was used by the church and, thereby, give more accurate calendars for planning church events. But the church became alarmed when the new theory removed the Earth and God's centerpiece, man, from the center of the universe. Instead of accepting, the church sought to defend its old dogma.

In 1543, Nicolas Copernicus published his *De revolutionibus orbium celestium* (*On the Revolutions of the Celestial Spheres*), a theory that moved the Earth from the center of the universe into an orbit around the Sun, where it became just one more planet. He established the order of planets and devised a system that accounted for planetary movement with uniform circular orbits. Later, Copernicus' circular orbits were corrected to be elliptical orbits by the astronomer Kepler, who used the detailed observations of Tyco Brahe to formulate his theories. After Copernicus, numerous others, including Brahe, Kepler, Galileo, and particularly Newton, extended the reach of the physics of Copernicus' theory with new laws that continue to be useful to this day.

Sixty years after Copernicus' theory, only a dozen or so scientists supported it, one of which was Galileo, the most famous scientist at the time in Italy. Galileo constructed a telescope in 1608 and directly observed the planets, Sun, and moon for the first time, discovered four moons around the planet Jupiter and spots on the sun. All of these observations conflicted with Vatican's dogma. Led by observations made by Galileo, Brahe, and others, scientists were able to peel away the dogma on the workings of the heavens that conflicted with observations.

Copernicus' helio-centric theory opened the door for a fresh look at the physics of the movements of the Earth, the Sun, and the planets by moving the Earth and man, the centerpiece of God's creation, away from the center of the cosmos. For the first time, a natural theory based on man's observations of Nature and codified by a godless natural theory had challenged God's supernatural plan enshrined in church dogma—and Nature won. One of Nature's secular theories had replaced one of the church's dogmas, the geo-centric theotheory.

The Vatican defended its dogma of a geo-centric universe by bring Galileo to trial for his observations and writings and as expected found him guilty. But Copernicus' new thoughts had already spread in Europe. After Galileo's trial, the scientific theories of Copernicus and Kepler found a receptive audience among other scientists, and science advanced on a widening arc of scientists throughout Europe.

In England, Newton, with his theory of gravitation took the next step in establishing Nature's theories describing the universe. He showed that forces on the Earth, such as the one that dropped an apple on his head, were the same as those that caused the motions of the planets in the heavens. His theory of gravitation was the accepted theory for planetary motion until Einstein's Theory of Relativity was introduced 340 years later.

CHURCH RESPONSE

The Vatican took Galileo's book defending Copernicus' Theory as an attack on its dogma (which it was) and put Galileo on trial for heresy. His crime was confronting the church dogma by supporting Copernicus with arguments that the Earth went around the sun, a theory confirmed by observations with his own telescope. Standing before the Vatican's inquisitors at his trial, Galileo argued in his defense:

> *I do not feel obliged to believe that the same God who has endowed us with senses, reason and intellect has intended to forgo their use . . He would not require us to deny sense and reason in physical matters, which are set before our own eyes and minds by direct experience or necessary demonstrations.*

It was a reasoned argument based on the nature of science and religion. Further, Galileo argued that scientific discoveries should be heard on the merits of science rather than religious dogma, for science was an independent discipline and the two had different aims:

> *The aim of Scripture is not to teach science, but to teach salvation.*

Later, Kepler also argued that the Holy Scriptures spoke colloqui-ally and poetically about common things, such as the Sun's apparent motion through the sky, "concerning which it is not their purpose to instruct humanity." Given the Bible's emphasis on salvation, Kepler advised believers to:

> . . . regard the Holy Spirit as a divine messenger, and refrain from wantonly dragging him into physics classes.

But the church did not listen to Galileo or Kepler and defended its dogma by asserting its authority over science and declared Galileo's support of the helio-centric theory to be heretical.

The Vatican's confrontation with Galileo in 1633 is called one of the most famous trials in all of science. And rightly so, for it was done in a grand style—a regal pope in purple and white robes surrounded by a jury of crimson-clad cardinals sitting in judgment over a lone, famous, rumpled, old scientist charged with arguing for a new physics theory that challenged the powerful church dogma. But the trial was not about physics, but about theocratic power. It was a classic example of the church defending the authority of its dogma over physics. Galileo's reasoning failed to convince the pope, and the Vatican court found him guilty of heresy (a religious crime). In the trial the science theory (Copernicus' Theory) was not judged on its scientific merits, for the Vatican clergy shifted the focus of the trial to be about defending their religious dogma.

Although the Vatican found Galileo guilty, Copernicus' theory has been accepted as a major step toward understanding the cosmos by the scientific community. There is a bit of irony in the Copernican Revolution, for the Vatican at first had urged a scientist, Copernicus, to help settle the problems of the Julian calendar in accurately setting dates for church affairs. To do this, scientists had to study the motions of the celestial bodies on which the calendar was based. Copernicus did, and his theory was the result.

GALILEO REVISITED

The Vatican has had a difficult time accepting that it had unjustly accused Galileo, and as late as 1990, Cardinal Ratzinger, now Pope Benedict XVI, cited views on the Galileo affair by quoting the philosopher Feyerabend, which created a stir with the scientific community:

> *The Church at the time of Galileo kept much more closely to reason than did Galileo himself, and she took into consideration the ethical and social consequences of Galileo's teaching too. Her verdict against Galileo was rational and just, and the revision of this verdict can be justified only on the grounds of what is politically opportune.*

Needless to say, many scientists disagreed with this misplaced defense by the Vatican of its sixteenth century dogma.

At the time Church sermons darkly predicted that God would bring disasters to those who removed God from controlling the cosmos and the Earth. But Copernicus' theory was accepted by the leading scientists, and the church dogma of the sun orbiting the earth was proven wrong. God has not sent doom to the Earth for removing the Earth as the center of the universe. The result has only been the church losing its godly authority over the movements of the celestial bodies in the natural world to Nature's authority.

During Galileo's trial, the Vatican did not attempt to resolve the technical validity of the claims about the orbit of Earth, but only to defend its religious dogma. The charge was that the helio-centric theory was heresy, and Galileo was convicted as being a "vehemently suspect of heresy" in his trial. Galileo's book was placed on the Catholic Index, a list of forbidden books, and for the next 125 years the church prohibited books presenting the physics of helio-centrism. In 1758, the church dropped the prohibition, and in 1822, the pope allowed the printing of helio-centric books. It was not until 1992 that the Vatican officially said that the church had come to the wrong verdict in the Galileo trial. Taking 359 years to correct a view on science is quite a record for any bureaucracy. Churches had great reluctance in giving up their theocratic control over natural science.

LUTHER

Both sides of the Protestant Reformation held comparable views on God's authority over science, for it was not only the Catholics who fought to keep the Christian earth-centric dogma intact, but also the Protestants. Martin Luther in 1539 took note of Copernicus:

> *There is talk of a new astrologer who wants to prove that the earth moves and goes around instead of the sky, the Sun, the moon...The fool wants to turn the whole art of astronomy upside-down. However, as Holy Scripture tells us, so did Joshua bid the sun to stand still and not the earth. This fool (Copernicus) wishes to reverse the entire science of astronomy, but the sacred Scripture tells us that Joshua bade the sun to stand still and not the Earth?"*

Catholics broadly condemned it, and in 1616 the priest Francesco Ingoli wrote that Copernicanism was:

> *Philosophically untenable and theologically heretical.*

Contrary to the Christian view, others, such as Delmedigo, a Jewish rabbi, was noted as saying that the arguments for Copernicus are so strong that only an imbecile will not accept them. Delmedigo had studied at Padua and was acquainted with Galileo, but he was outside of the powerful Catholic organization, and his words had little effect.

The removal of the Earth from the center of the universe to a lesser role in the solar system by Nature's godless physical laws of planetary motion has not proven to be a frightful event as predicted by many in the church; the skies have not fallen, nor have the churches. Few people now question the removal of God and the installation of Nature as the authority over celestial mechanics.

LATER

About four hundred years after Galileo's encounter with a pope, another encounter took place between a scientist and a pope, Stephen Hawking and John Paul II. The encounter was at a scientific conference held in the Vatican where Hawking was to deliver a scientific paper on the creation of black holes. He accidently met the pope, who in passing offered Hawking some unsolicited papal guidance:

> It's OK to study the universe and where it began. But we should not inquire into the beginning itself because that was the moment of creation and the work of God.

Hawking jokingly told the pope that he would be giving a paper at the conference on how the universe began[16] and suggested that the pope should give one too. The fact that the pope thought that he could inject religious guidance into science and provide limitations to the range of science ("we should not inquire into...") indicated that the Vatican still held the belief that their religion had the authority to control the extent of science. The power of the Vatican over science is long gone, but the encounter, however mild, is a reflection of difficulties the Vatican has had accepting science as being independent of religion. Hawking's reply was a polite rejection of the pope's suggestion of a limitation on science while extending to him the same rights his fellow attendees had at the conference—the right to present a scientific paper for review by the secular scientific community[17]. Both science and religion can inquire into the birth of our universe, and can each have a theory: the pope's theotheory based on the supernatural Christian narrative and Hawking's scientific theory based on natural science observations.

Since that time Hawking has continued his work on the origins of the universe. In his recent book, *The Grand Design*, Hawking gives a summary hypothesis on how our universe came into being from a small, super dense singularity, which could have been titled *Thank Nature for the Universe*. He argues that theoretical physics theories (interactions of space-time and energy) give us insights into the creation and nature of our universe. One of Hawking's insights is the role of time, or specifically the creation of space-time in a quantum fluctuation, which is proposed to have initiated the Big Bang. The absence of space-time before the Big Bang precludes an in universe cause. The pope is probably not happy with Hawking's theory.

With science, man has been able to acquire sufficient knowledge to build telescopes to peek at more and more of the universe. The deeper in space an object is observed, the further back in time it is. Telescopes now allow us to see back over thirteen billion years at almost the moment of the birth of the universe. Observations such as the mappings of the cosmic microwave background radiations left over from the Big Bang have presented us with a view of the early universe. From these observations we are beginning to learn about the physics and workings of chance in the universe at times almost back to the Big Bang (about 380,000 years afterward). As the physicists say, "We are now beginning to see the very fingerprint of Nature, or as some say of God."

Experiments on understanding the universe continue with other tools, such as the Large Hadron Accelerator at CERN, which search for particles that may have existed fractions of a second after the Big Bang. The discovery of one such particle, the Higgs, was important to physics in confirming in more detail our understanding of fundamental particles. Other experiments continue on understanding dark energy and dark matter, both of which we know little about.

All of this is contrary to the pope's suggestion on what science should not do—try to understand the creation of our universe. Nuclear particle accelerators give us fleeting peeks from high-speed collisions at the small specks of matter and energy that are the building blocks of the universe. Three of these little specks, or quarks (there are six different ones) and electrons combine and interact to constitute the building blocks of matter we see in the world around us. There may be others for us to discover.

On the smallest of scales, we also see chance at work in Nature. The basic interactions at the subatomic level described by quantum physics are based on probabilities, not certainties, of events occurring. Nuclear particles such as the neutron, and even elements we see in Nature, such as uranium, exhibit an uncertainty about the longevity of their existence. Some are radioactive and probabilistically decay from one element to another to another over time; in the case of uranium decays into lead.

Verification of our understanding of Nature is based on these observations. In physics there is the Standard Model, an overarching theory describing all that is known about matter and energy at this time. But much remains unknown, such as what dark energy and dark matter are and how to fit gravity into the grand scheme of the physics model. String theory is offering hope in putting all of the pieces together, but without a way to test it so far, many questions remain.

Even now, over four hundred years later, Nature's authority over the creation and design of the universe is still being challenged. To keep God's authority over the universe, some Christians today argue that they can see the proof of God in the "fine-tuned" universe around us. They argue that the values of fundamental physical constants of particles, the force of gravity, nuclear forces, mass of the proton, etc., would not support life today if they were not exactly what they are. This is called the anthropic principle, a favorite argument of John Polkinghorne, an Anglican priest and scientist[59], who proposes that the universe must have been designed by God to be so well tuned. Scientists argue that Polkinghorne's view is just another example of the supernatural Intelligent Design theotheory, in this case applied to the evolution of the universe. Rev. Paley, who two hundred years ago argued that the complexity of biological life required a designer, must be proud of Polkinghorne who uses his argument for the need of a God to design the universe.

DEMOCRATIC REVOLUTION

Two hundred and fifty years after the Copernican Revolution, the next intellectual revolution impacting religion was about governance, the Democratic Revolution, and it happened in America. Selected citizens became members of the Continental Congress and charged with the task of forming a democracy. This resulted in the writing of a secular constitution which was based on the authority of the people. It did not include God's authority over the state even though it was written by a convention composed of a majority of Christians in 1787.

Two years later, after ratification by the states, our country began governing under its secular laws. Fears expressed by some Fundamentalist Christians that the secular Constitution, omitting God's guidance over the country, would cause a breakdown in morals and a decline in religious worship in the country, have not been realized.

Our country has not become immoral by separating God from the State. In fact, one can argue that the opposite has occurred, for by accepting secular morality for the bases for the laws, our country's moral standards are an improvement over biblical laws embedded in the Christian scriptures which included slavery and little rights for women. Although it has taken time slavery was abolished by the United States in 1864 during the Civil War and women's rights took another sixty years. These changes in governance under our secular Constitution have not impacted the Christian God's message to believers, and Christianity is as vibrant today as it was at the founding of our country, if not more so.

Today the scientific community, arguing for Nature's rights, should be attempting to get the same contract with our democracy that was granted to religion—that it be given a separate and independent status from the government and religion. Such a contract for science would mean building a wall separating God's authority from Nature's (science) authority and giving scientific authority to the scientific community (the People). The continuing conflicts with religion over science (for example, Darwinian evolution) in school textbooks would be eliminated. Questions on scientific content would be monitored by the scientific community rather than by political and religious groups that can be (and have been) guided by religious members who wish to inject their God into science. As the Copernican Revolution has shown, God's science presented by religious dogmas is not as helpful to mankind as Nature's science presented by the scientific community.

CHURCH RESPONSE

Since the establishment of our secular Constitution, there have been periodic appearances of unsuccessful movements to incorporate the

Christian God into the government and make the country a "Christian Nation" (which it has never been). Minor symbolic changes in the inclusion of God into governmental trappings were made during the height of the Cold War in the 1950s and 60s, such as inserting "under God" into the Pledge of Allegiance and putting the phrase on paper money in order to differentiate the country from our Cold War adversary, who we declared to be Atheists. This was voted in at a time of great stress for the country from a potential nuclear war with a foreign power. Even after the threat of a nuclear war diminished, some Fundamental Christians continued to pursue their concept of a "Christian Nation" and used phrases like "return America to its Christian roots," even though in doing so, they showed a profound ignorance of American history.

The history of the formation of our Constitution is clear. The Continental Congress charged with writing the Constitution gave Christian delegates adequate time to make their case for a "Christian Nation." Motions to have the Congress insert references to God in the Constitution (example, a preamble with God included) were heard and rejected. The Founding Fathers clearly did not wish our country to be a Christian nation and instead voted for a secular State defined by a secular Constitution that did not reference God.

PEDOPHILE CRISIS

The pedophile crisis within the Catholic Church is not a belief or Christian issue; it is a State governance issue illustrating the difficulty churches have coming to grips with modernity and separating the church from the State. This is true in the US as it is in other countries. Instead of embracing the government to enforce the secular laws of the country, as all citizens and organizations must do, the Catholic Church tried to ignore the civil laws and define justice by the church bureaucracy. They failed miserably.

The exposure of an ongoing worldwide pedophile crisis has unveiled a sad legacy of a culture of religious management secrecy, which allows the church and the clergy to be protected at the expense of the civil rights of the congregants guaranteed by the secular Constitution. When crimes are committed against the People, who in this case were mostly children molested by the clergy, the Churches' first actions have been to protect the clergy and the organization, not the victims, the children. The church's secrecy and king-like management, along with little clergy accountability to state laws, have proven to be an immoral environment that did not protect the rights of a powerless laity, including children.

The handling of pedophilia by the Catholic Church hierarchy has revealed the management backwardness and "we-are-above the secular law" mentality of the church leaders. Even responding to a public call for reform, the Vatican has done poorly, for it simply does not know how to share oversight with the community for management of its churches so that the congregants (the People) are treated fairly and protected to the standards of the secular society. The church has never had to, so it never learned. It is not able to treat the State as the independent discipline to enforce the state laws. Author Maureen Dowd commented[60] on the Vatican's document on improvement in church policy when it was obvious that it failed to pass the standards of secular organizations:

> *The casuistic document did not issue a zero-tolerance policy to defrock priests after they are found guilty of pedophilia; it did not order bishops to report every instance of abuse to the police; it did not set up sanctions on bishops who sweep abuse under the rectory rug; it did not eliminate the statute of limitations for abused children; it did not tell bishops to stop lobbying legislatures to prevent child-abuse laws from being toughened.*

The rights of victims, children in Catholic churches, have been poorly served by the Vatican. It has become clear that secular laws employed by secular organizations outside of the walls of the church provide a higher level of morality in this case. The Vatican is simply not capable of understanding the governance of moral issues with their congregants to the standards of the civil society.

It is not only the church that is at fault, but the community at large for accepting the false assumption that religious organizations are different—that they can govern themselves to the laws of the land and that their organizations are morally superior to secular organizations. In effect, governments have essentially abdicated their civil responsibility and authority to ensure safety of its citizens to religious organizations through lack of oversight.

The reality is, of course, that churches are not and never have been any better or worse than comparable secular organizations. However, they have proven to be among the worst at civil governance, as has been demonstrated by the size and depth of the widespread pedophile crisis, which they have allowed to fester. Secular oversight over church leadership and transparency in their operations are, as with any other organization, necessary to protect the public.

EDUCATION

Our democracy has as one of its strengths the commitment to a secular education for all of its citizens. Citizens of all religions are welcomed. Secular public education supplied by the secular State broadens the acceptance of natural science and increases religious tolerance by introducing Christians (and all believers) to members of other religions.

Teaching natural science is a democratic mandate for public schools, which, unfortunately, is not being followed throughout the country. A recent study[61] has found that in our public schools, only 28 percent of biology teachers follow the scientific recommendations of the National Research Council on teaching Darwin's Theory and its underlying role as the unifying theme of all biology. About 12 percent explicitly teach a religious theotheory (Christian Creationism), while 60 percent attempt to avoid controversy by endorsing both. By avoiding intellectual commitment to Darwin's Theory, a major foundation for scientific literacy is missing. In effect, Christian pressure on the state's public educational system to argue for God's authority over Nature and State is "dumbing down" our students.

In some regions of the country, Christian schools provide not only religious instruction, but general education. The good news is that it is always good to have more educational facilities; the bad news is that sound natural science instructions are not being accommodated in smaller parochial schools, thereby placing those students at a disadvantage relative to those educated in larger secular school systems.

FEAR OF SCIENCE

There is concern, even a fear by some Christians, that having Nature replace God as the authority for natural science to explain man's evolution and the workings of the natural world, would turn our society into an amoral one, empower Atheists at the expense of Christians, and lead to science becoming a religion.

These are unfounded fears. Science is neutral with respect to religion and provides no support for or against religious belief. Science is universal, which requires it to be godless. Christians, Buddhists, and Hindus (and believers of all other religions) can use science without fear of impacting their gods or religions in the supernatural world. Natural science has the authority to explore only in the natural world. Science has no religious position; it deals only with natural world hypotheses and theories, which are continuously being tested and changed when in error.

There is little to fear that science will become a religion. The scientific process is self-inoculating against supernatural religion; it cannot go there, for it does not address the supernatural. Conversely, religion does not need the rigors of science to do its principal work—support humanitarianism.

DARWINIAN REVOLUTION

Some three hundred years after Copernicus, the naturalist Charles Darwin started a new scientific revolution—one that would bring biology, the science of living organisms, into Nature's tool chest alongside physics and chemistry, the tools that had been used in the Copernican Revolution. Darwin's Theory of natural selection is an explanation of the evolution of all living organisms, including man. It removed God from being the creator and designer of living organisms and replaced him with Nature.

It was Darwin's genius to discover that it was Nature, a natural process, and not God, a supernatural force, that was the designer "who created and designed us." Darwin's great idea was to assign Mother Nature as the authority over all living organisms, including man. As Ayala notes:

> It was Darwin's greatest accomplishment to show that the complex organization and functionality of living beings can be explained as the result of a natural process—natural selection—without any need to resort to a Creator or other external agent. The origin and adaptations of organisms in their profusion and wondrous variations were thus brought into the domain of science.

Three billion years ago or so, life, in the form of the first self-replicating cell, appeared on planet Earth, and from that time to now the evolution of living organisms has proceeded according to Nature's laws without the need for a supernatural God. From that initial common ancestor, evolution has been at work slowly, in small steps, according to the process described by Darwin's Theory and has produced living organisms with the great diversity, including modern man, we see around us.

Our tree of ancestors has many branches, but there are no immutable categories of animals or even hominins requiring new life forms. There are no boundaries that separate plants from animals or from humans on our evolutionary tree; only one tree of life with a tangle of branches from which by chance the next species evolved, while others were selected for death. This is nature's process, which has been repeated many times over with one of its last survivors being man.

Chance plays a critical role in natural selection, for with random changes in DNA within a species, natural selection determines the survivors. Chance is inherent in the micro-biological and the macro-level with environmental changes that impact natural selection. An example on the macro stage is the chance hit by an asteroid on the Earth sixty-five million years ago that resulted in the deaths of many species, including most of the dinosaurs. Without that "chance happening" a mousey little animal living in the shadows of the mighty dinosaurs might still be in the shadows. But it did hit, and the mousey little animal survived, endured many other encounters, and over time evolved into larger animals, on to monkeys, to the great apes, and to man.

In our evolutionary tree, there have been many millions of Adams and Eves (at each species branch point), each contributing a little biological change to the species, some of which were man's ancestors. In one branch of the great apes, about eight million years ago, some of our ancestors came down from the trees, adopted bipedalism and for reasons not yet totally understood their brains made a spurt of growth in size from that of chimps to that of man, about three times larger. During this evolutionary period about two million years ago, our ancestors invented tools that gave them a higher probability of surviving by becoming more efficient food providers. Further, effective speech appeared which opened the door for increased socialization and mental capabilities. These early pre-hominins evolved into several species, one in our branch of the tree with others existing in parallel: the Neanderthals and the Hobbits, a smaller species of hominins.

Over this last two million years, our ancestors[25], the hominins, appeared in our family tree with increased brain sizes. A larger brain and the capability for speech are supportive explanations for an increased mental capacity, which accelerated the expansion of more complex social behavior. Finally, about a hundred thousand years ago, Homo sapiens, our species, arrived on the scene. Each of the incremental Adams and Eves in our long ancestral lineage is as important as the previous one, for in each case their offspring survived and contributed a little change to the next generation on our path through the tree of life.

The feedback between increasing tool usage and increasing brain capacity appears to have been autocatalytic for increasing mental development—that is, the brainpower to develop tools allowed us to do more, and the more we did with the tools, the greater the possibilities that we could do new things, including make new and more efficient tools. The time at which speech was available to hominins is not known, but increased communications could have led to increased socialization and provided the mechanisms of indirect reciprocity and group selection, both of which have apparently played an important role in human evolution.

The Darwinian Revolution is far from being over and the study of the evolution of the brain-to-mind is well underway. The evolutionary biologist Ayala highlighted two important research challenges for evolutionary biology and the emerging neuroscience revolution:

> *Human biology in the twenty-first century faces two great research challenges: The ape-to-human and the brain-to-mind transformations. By the ape-to-human transformation, I refer to the mystery of how a particular ape linage became a hominini linage, from which it emerged, after only a few million years, humans able to think and love, who have developed complex societies and who uphold ethical, aesthetic and religious values, by the brain-mind transformation.*

An understanding of the biological evolution of man and the diversity of living organisms has been provided by the Darwinian Revolution. New neuroscience discoveries have added to our understanding of the biological base for the evolution of man's moral sense and mental capabilities and have ushered in a new scientific revolution exploring the brain-mind transformation.

DARWIN'S STAGE

The scientific stage that Darwin encountered when starting his research in the early 1800s began a hundred years before, when discoveries of fossils and rocks begin to raise questions of man's origin and the age of the Earth. The discipline of geology was just beginning, and the rocks being explored indicated that the age of the Earth could be far older than the biblical description of six thousand years, possibly even millions of years or more. Fossil bones were being unearthed that raised questions on how the great diversity of life could have appeared in Nature and indeed how all of life is related. Neanderthal fossils were discovered first in 1829.

Early in the 1800s, William Paley (1743-1805), a clergyman, published *Natural Theology or, Evidences of the Existence and Attributes of the Deity Collected from the Appearances of Nature*, which argued that the complex design of organisms of life could not have come about by chance or by known laws of physics and chemistry, but had to be accomplished by an intelligent designer (God). Since Paley's days this argument has been a recurring theme for Christian explanations for the evolution of man.

Also early in the early 1800s, several concepts of evolution were being discussed in scientific circles. One of those thinking on the subject was Alfred Russel Wallace, who independently envisioned natural selection as promoting evolution about the same time as Darwin, but he did not link it to the design of living organisms. But it was to be Charles Darwin's genius to resolve the question, "who created and designed us?" As Ayala notes:

> *It was Darwin's greatest accomplishment to show that the complex organization and functionality of living beings can be explained as the result of a natural process—natural selection—without any need to resort to a Creator or other external agent. The origin and adaptations of organisms in their profusion and wondrous variations were thus brought into the domain of science.*

Darwin's great idea was to assign Mother Nature as the authority over all living organisms, including man. To do so he had to separate Nature's processes from God's supernatural world.

When published in 1859, Darwin's Theory lacked a satisfactory theory of inheritance, and the mechanics of natural selection were open to question. Mendel's theory of genetics (how genes pass on characteristics between generations), although written during Darwin's time, was unknown by him. The mechanism of Mendelian inherence was not used to support Darwin's theory until the early part of the twentieth century, when in 1930 it was incorporated into the theory. This revision of Darwin's Theory is known as the "modern synthesis."

Darwin knew that his theory of natural selection did not require intervention by God and that this would cause his wife and her religious community concern. For some Christians it would be more than concern, it would be heresy. But even if his theory was declared heresy, by the mid-1800s the Christian churches in England had little power to interfere or have the government interfere, so Darwin could let the science chips fall where they may. England of the mid-1800s was not the Italy of the 1600s, where the Church brought Galileo to trial for a heresy far less controversial.

But Darwin was sensitive to his wife's Christian beliefs, and for this reason, plus his natural inclination to gather more information, he delayed publishing his work for almost twenty years. Wallace's emerging thoughts on natural selection were similar to Darwin's, but he differed by insisting that our species was unique, while Darwin did not. Pushed by Wallace, Darwin published his book *On the Origin of the Species*. Darwin clearly establishes natural selection as the designer of man, thereby removing God as the designer. He introduced chance as a key factor and linked evolution to a common ancestor of all living organisms. Darwin noted:

Man still bears in his bodily form the indelible stamp of his lowly origin.

Charles Darwin was interested not only in physical evolution, but also in the evolution of morality and how it fits the human-animal continuum. Darwin was able to move beyond his contemporaries and open up new stages for studying biological and moral evolution.

TODAY'S STAGE

Subsequent support for Darwin has been provided by the discovery of the DNA molecule and its double helix structure in the mid-twentieth century and by the decoding of man's whole genome, a project completed at the end of the twentieth century.

The fossil record continues to grow, example over 400 Neanderthals have been found, and much sought after transitional species have been discovered. One outstanding example is the transitional fossil, Tiktaalik[20],an amphibian, which graphically illustrates the structural changes that were made in the transition from our fish ancestor to land animals 265 million years ago. Closer to us in the tree of life are the fossils of primates, which appeared 50 million years ago. This primate branch of the tree includes apes, chimps, and hominins, one of which is man. A recent and most interesting hominin[21] fossil find was only three feet tall. Named the Hobbits by scientists, they were found on the island of Flores in Indonesia. Radio-dating indicates that the Hobbits were still living, possibly to about ten thousand years ago, and illustrate that our already bushy parallel evolutionary branches could be even more diverse than we have thought.

Darwin would have been most happy to have his weak argument on heredity bolstered by Mendel's robust theory of genetics and by DNA's digital language of evolutionary genetics, thereby rounding out the completeness of his theory. Detailed support of Darwin with Mendel's genetics has come from the discovery of the DNA molecule and the expanding field of DNA analysis has greatly improved our understanding of the biology of life. The expanding capability to decode the DNA of a genome has given the evolutionary biologist the tools to examine the evolution of past and present living organisms at the molecular level.

An example of this is given in Appendix D: The Evolution of Vertebrate Blood Clotting, which gives a summary of the evolution of this process in many species over the last five hundred million years.

With the discovery of DNA, the power to understand biology at a fundamental level has been revealed. As Paul Churchland notes:

> *With the physical structure of DNA finally made clear, its all-important functional properties were slowly but steadily being revealed by chemical research. A purely materialist, reductionist account of the nature of life—of self-replication, of genetic diversity, of evolution of protein synthesis, of developmental and metabolic regulation—seemed to many scientists to be all but in hand.*

In his book *Out of Eden*, Richard Dawkins quotes A. E. Houseman's poem on the power of Nature's DNA:

> *For Nature, heartless, witless Nature,*
>
> *Will neither know or care,*
>
> *DNA neither knows or cares.*
>
> *DNA just is. And we dance to its music.*

As the poem notes, DNA neither knows nor cares; it is the godless evolutionary language of life for man, beast, pond scum, and virus.

There have been many other contributors to extending Darwin's Theory beyond its original scope. Lynn Margulis has advanced a supporting theory of biological change, called the endosymbiotic theory[18], which suggests a degree of cooperation between organisms that enhances survival chances. At the cellular level, more complex forms of life could arise from the merger of simpler forms with one another. Each simple form of life supplied some of the ingredients for the modification of something new.

Further, other impacts on species survival are being found. Epigenetics describes added evolutionary pressure from the bi-directional interchange between heredity and the environment, and sociobiological factors impacting species survival.

DARWIN'S THEORY

It was Darwin's Theory that tied all of the pieces of the science puzzle of man's evolution together: natural selection process, common ancestor, species variation, chance, and design sketched an overall answer to the scientific puzzle of the great diversity of living organisms.

In his publication of *On the Origin of the Species*[19] in 1859, Darwin sets forth his thoughts on natural selection in meticulous detail. Ayala briefly summarizes Darwin's Theory:

> *The theory of evolution conveys chance and necessity, randomness and determinism, jointly enmeshed in the stuff of life. This was Darwin's fundamental discovery, that there is a process that is creative, although not conscious.*

Darwin's Theory of natural selection accounts for the "design" of organisms, and for their wondrous diversity, as the result of natural processes, the gradual accumulation of spontaneously arisen variations (mutations) sorted out by natural selection. Which characteristics will be selected depends on which variations happen to be present at a given time in a given place and survives. This in turn depends on the random process of mutation as well as on the previous history of the organisms. Mutation and selection have jointly driven the marvelous process that, starting from microscopic organisms, has yielded orchids, birds, and humans. The theory of evolution conveys chance, necessity, and randomness, jointly enmeshed in the stuff of life. This was Darwin's fundamental discovery: that there is a process that is creative although not conscious.

Darwin's Theory is in stark conflict with God's biblical creation story. God's certain creation of man was being challenged by a natural process, which used variations by mutations in a species and subjects them to survival by natural selection. Ayala notes that the process is creative and produced the design of a living organism:

> . . . including humans who think and love, endowed with free will and creative powers, and able to analyze the process of evolution itself that brought them into existence.

This powerful explanatory and predictive theory of natural evolution has become the central organizing principle of modern biology and provides a unifying explanation for the diversity of life on Earth.

The process of natural selection is conceptually straightforward, even though living organisms are complicated machinery with many genes that have interactions and expressions. Natural selection involves random gene variation within a species that undergoes natural selection, a process that passes adaptive changes in an organism's genes that survives to the next generation. It is a process that keeps repeating for each generation which must survive in a randomly changing environment.

Table A: Key Tenets of Darwin's Theory	
1. Natural process	Natural selection is a process occurring in Nature without supernatural influences.
2. Long time frame	Evolution of life on Earth has occurred over several billions of years, possibly 3.5 billion years.
3. Common ancestor	All life has descended from a common ancestor from which the great diversity of life has emerged.
4. Design by Nature	All living creatures are designed without a designer as a result of Nature's process of natural selection.
5. Random Mutations	Random variations occur within a species from random mutations.
6. Chance Outcomes	Evolution is the result of natural selection of random mutations within random environments

Table A lists six key tenets of Darwin's theory. They involve the process of natural selection operating in Nature over a long time frame of billions of years from a common ancestor. The design of living organisms is without a designer, for it is by Nature's process, which includes chance.

NATURAL WORLD

Nature's theories describe natural process in the natural world. One of these is Darwin's Theory, the evolution of living organisms, a natural process using only Nature's tools (physics, chemistry, and biology).

Darwin's Theory leaves the supernatural biblical creation theotheory valid only for religious studies in the church. As Karen Armstrong notes:

Evolution (Darwin) has indeed dealt a blow to the idea of a benign creator, literally conceived. It tells us that there is no Intelligence controlling the cosmos, and that life itself is the result of a blind process of natural selection, in which innumerable species failed to survive. The fossil record reveals a natural history of pain, death and racial extinction, so if there was a divine plan, it was cruel, callously prodigal and wasteful. Human beings were not the pinnacle of a purposeful creation; like everything else, they evolved by trial and error and God had no direct hand in their making.

Ayala agrees that Darwin's Theory of natural selection is important, but argues that Darwin's most fundamental idea is that the design of life arises from natural processes. Ayala notes:

This is Darwin's fundamental discovery, that there is a process that is creative though not conscious. And this is the conceptual revolution that Darwin completed: the idea that the design of living organisms can be accounted for as the result of natural processes governed by natural laws.

The natural processes used in the creation, evolution, and design of living organisms use only Nature's tools (physics, chemistry, and biology), all of which embrace chance. This departs completely from the Christian theological view expressed in the theotheories that the design of man is by the direction of a supernatural God, contains no uncertainty, and employs miracles.

Many fossil discoveries have given physical evidence to support Darwinian natural selection and the diversity of life. In addition to the fossil evidence, Ayala outlines the in-depth support for Darwin's idea from the molecular biological research in which he is engaged:

Molecular biology has made it possible to reconstruct the "universal tree of life," the continuity of succession from the original forms of life to every species now living on Earth.

Fellow evolutionary biologist Russell Doolittle has made a reconstruction of a part of the tree of life for the blood clotting system over the last five hundred million years (see Appendix D). Molecular biology, Ayala states, provides strong evidence for the evolution of living organisms:

First, by showing the unity of life in the nature of DNA and the workings of organisms at the level of enzymes and other protein molecules, second, and most important for evolutionists by making it possible to reconstruct evolutionary relationships that were previously unknown, and to confirm, refine, and time all evolutionary relationships from the common ancestor up to all living organisms.

In effect, Ayala confirms Darwin's Theory from his work as a molecular biologist. He gives the authority of natural selection to Nature, as did Darwin.

Long Time Frame

Nature's time schedule for the evolution of living organisms is long—billions of years, for genes accumulate random changes from many small mutations over many years and are the source material for evolution's natural selection. The span of time covering the evolution of life on earth has been determined from dating rock strata of ancient fossils to be over several billion years. Present data indicates that life began on Earth about 3.5 billion years ago and has proceeded from a first common ancestor to the wild diversity of life we see today.

COMMON ANCESTOR

A basic tenet of Darwin's Theory is that all life came from a common ancestor. The composition of the first self-replicating molecule and the first cell are not known, but after that, all life has proceeded by Darwinian evolution for about 3.5 billion years. DNA provides a molecular trail back to the earliest life forms. Some of the fossils found have given DNA samples that have aided the positioning of fossils on the evolutionary tree of life.

> *The primordial organisms were very small and relatively simple, yet all living things have evolved from these lowly beginnings.*

Ayala states that biological research has progressed to the point that:

> *. . . gaps in the reconstruction of evolutionary history from all living organisms back to their common ancestor no longer exists.*

DESIGN

Darwin's Theory is a revolution in the biological sciences that answers the question of the design of living organisms. Ayala notes:

> *{Darwin's} idea {is} that the design of living organisms can be accounted for as the result of natural processes governed by natural laws.*

All living organisms are:

> *designed without a designer.*

Ayala stresses the importance of Darwin's Theory to the design of living organisms by noting that Alfred Russel Wallace, who was also thinking of evolution at the same time as Darwin, did not formulate the next step and explain natural selection's contribution to design:

Wallace saw natural selection as promoting evolution, which he saw as progressive, not as the explanation of design, which is its ultimate significance.

It was Darwin's great idea that natural evolution is a process that is creative but not conscious. Darwin's Theory achieves evolutionary design changes in species through the process of natural selection operating on random variations in the host species. The natural process results in biological design of all living organisms without a designer. Fish are "designed" by natural selection with attributes that give them a better chance of surviving in a watery environment, such as hydrodynamically efficient bodies than allow faster speeds and less energy output. The same for birds, but in an air environment where consideration of wing design is critical for speed, lift, and energy output.

It was Darwin's greatest accomplishment to show that the complex organization and functionality of living beings can be achieved as the result of a simple natural process—natural selection. Nature is creative; it creates without a creator and designs without a designer. There is no need to resort to a supernatural Creator to explain life's design or its diversity. Darwin brought the study of living organisms (life) into the domain of natural science and made them explainable with Nature's processes using Nature's tools: physics, chemistry, and biology.

The arguments that complex organisms must have a designer and a creator outside of Nature are no longer credible, although they continue to be used by many Christians. The complex blood-clotting sequence requiring two dozen different steps is often put forward as an example of a complex design that must have had an intelligent designer (God) to put the many steps required in clotting in place for it to work in man. The problem with this argument is that a scientist, Russell Doolittle, has worked out the detailed steps of the evolution of the complex string of sequential molecular steps required for blood clotting (see Appendix D: Overview of the Evolution of Blood Clotting). This physical explanation of Nature's evolved design should remove arguments for the need of an intelligent or innovative designer, for Nature has proven to be a very competent designer.

Other complex biological organs, such as the eye, have also been shown to be made by natural selection. Neil Shubin's book *Your Inner Fish: A Journey into the 3.5 billion-year History of the Human Body*, reveals many earlier evolutionary designs that can be found as the basis of man's evolved design today. In the natural world, Nature has been shown to be a competent designer. Rainbows and snowflakes are beautifully designed, but they, like man, are designed without a designer, for it is Nature's processes (physics, chemistry, and biology) that designed and produced them.

CHANCE

Chance plays a central role in how Nature works. The heavy elements in our bodies were formed in distant starts that exploded: they are cosmogenic, that is, of cosmic origin. The element, iron which is critical to our hemoglobin's transport of oxygen in our blood is of cosmoenic origin.

Our carbon 14, one of the isotopes used for dating fossils, is an interesting example of chance in Nature. It is also cosmogenic, formed by neutrons from cosmic rays reacting with nitrogen 14, the most common element in our atmosphere, to form C14, which mixes with oxygen to form carbon dioxide. C14 is radioactive, with a half-life of 5730 years. The intake of C14 stops when living organisms die, and the decaying ratio of C14 to the non-radioactive C12 gives an estimate of the time since death.

Chance is an integral part of Darwin's natural selection process. Even the unpredictability of the weather gets to contribute to the uncertainty of human evolution, as do random physical events such as volcanic eruptions and meteors hitting the earth (dinosaurs would agree with this). The coming and going of global ice ages and the resulting disruption of weather patterns appear to have impacted the evolution of pre-hominins and hominins in and out of Africa. It is estimated that man's branch of the evolutionary tree was probably down to a population of less than a thousand at one time. Man is just the last hominin standing after surviving much evolutionary turmoil.

Some Christians are concerned and will not accept chance being embedded into man's evolution, for chance directly confronts their belief in God's certain direction for the creation and evolution of man. But God's steady hand guiding Nature is not what science observes in the natural world. What science observes is that chance is fundamental to Nature's processes.

One then can appreciate the concern of the Christian authors about the inherent uncertainty with natural Darwinian evolution. One of the Christian authors notes that he was taught in science classes to find order and simplicity in the universe, for that approach was a guiding principle of physics. After all Einstein was able to make fundamental discoveries of Nature (Theory of Relativity) following this principle. But all of Nature does not run with the certainty that many, including Einstein, desired. With the emergence of quantum physics and its embracing of chance as being fundamental to Nature, Einstein was concerned with this view and argued against it. It is said that Einstein, in a discussion with Neils Bohr, a fellow physicist, argued that God[22] would not agree with the uncertainty in Nature demanded by quantum mechanics. In a famous quip to Bohr, Einstein said:

God does not play dice.

Bohr, in frustration, said, "Einstein, stop telling God what to do." Projecting what science theories should or should not do when they conflicted with observable data was not acceptable to Bohr.

As it has turned out, Bohr was right, and Einstein's view of what God or Nature wanted was wrong. Christians are equally as wrong telling their God what to do in the science of evolution, for Nature does not listen to the Christian God, the Hindu gods, or any other god, for it marches only to its own physics, chemistry, and biology.

Uncertainty is found in many places in Nature. A close look at how fluids flow has been applied to the dynamics of weather, providing a mathematical explanation for why it is not possible to make weather forecasts months in advance, nor planetary motion many years out, for there are infinitesimal uncertainties or errors in the information, which causes forecasts to diverge from reality. Mother Nature seems to enjoy uncertainty, for she uses it in so many places. The more science reveals the workings in biology, chemistry, and physics, the more that scientists discover that uncertainty is part of the fabric of Nature.

Now some seventy years after Einstein's quip about quantum mechanics' uncertainty, Nature's embrace of uncertainty continues to give us deep insights into the structure of elemental subunits of particles, to particle interactions and even to the Big Bang itself. Uncertainty may be uncomfortable and unsettling to some, but it is integral to our understanding of Nature. If one accepts that Darwinian evolution does embrace the inherent uncertainty in the random mutation of genes, DNA replication errors, gene shuffling, and natural selection, we have a better understanding of man's evolution and the great diversity of life we see around us. Nature with its uncertainty is what is observed from the very particles that makeup the universe.

Christians are left to answer the question:

Why would God make the outcomes of his creations riddled with randomness so that he cannot predict the outcome?

Some Christians counter that God, being omnipotent, can insert or remove uncertainty in his creations with miracles at his will. This is a difficult argument to defend, for why would an all-knowing presence need uncertainty (can God not predict the outcome?) on a path if he knows for certain where the path ends? The answer for Christians is simple: God's narrative in the supernatural world of religion does not have to embrace uncertainty, but Nature in the natural world does.

As a biological researcher, Ayala is well aware of evolutionary process and recognizes the role that chance plays in it. The description of the role that chance plays in natural selection was proposed by Darwin:

> *Chance is, nevertheless, an integral part of the evolutionary process. Mutations that yield the hereditary variations available to natural selection arise at random. Mutations are random or chance events.*

Additionally:

> *The theory of evolution conveys chance ... randomness and determinism interlocked in a natural process that has spouted the most complex, diverse, and beautiful entities in the universe.*

Those Christians who wish to remove chance from evolution and make evolution God-centric in the natural world will get no support from natural selection's dependence on chance. Understanding the role of chance eliminates the old theotheory of God's intelligent design of life by William Paley, a concept that has been used by Christians for a century and a half to argue that only God could have created the complex designs of animals and man.

MORAL SENSE

Charles Darwin viewed the evolution of morals as a result of natural selection—that is, morals are part of the biological evolution of living organisms. He placed morality in context with the animal-human evolutionary continuum. In *The Descent of Man and Selection in Relation to Sex*, Darwin noted that:

> *any animal whatever, endowed with well-marked social instincts...would inevitably acquire a moral sense or conscience, as soon as its intellectual powers had become as well developed... as in man.*

Other philosophers and scientists also argue that man's morals are a result of his naturally evolved (biologically based) moral sense. Jerry Coyne, a biologist, summarizes philosophical arguments on the origins of secular morals from evolved creatures:

> *Beginning with Plato, philosophers have argued convincingly that our ethics come not from religion, but from a secular morality that develops in intelligent, socially interacting creatures, and is simply inserted into religion for convenient citation.*

Philosopher, H. Scott Hestevold[23] has addressed the question of what the relationship is between morals and God and whether morals can exist without a God. He concluded:

> *If, however, such moral facts do exist, there is good evidence (a) that God did not cause the moral facts to be what they are, (b) that one's having knowledge of moral facts does not depend on one's having knowledge of God's moral views, and (c) that belief in God is neither a necessary nor sufficient condition for fulfilling one's moral obligations.*

For Darwin, his theory solved a major conflict he had with the scriptures:

How could an omnipotent, beneficent god design creatures that did so much harm?

It was Nature's process of natural selection that was responsible for the deaths in the natural world and not god from his supernatural world. Peoples have long lived with Nature's fury that brought deaths—lightning, storms, plagues, diseases, spontaneous abortions, volcanoes, etc. If that's caused by God, then he is far from the kindly, loving, peaceful God Christians preach about. Ayala notes that 20 percent of all recognized pregnancies end in spontaneous abortions, a fact that means a woman's body's is poorly designed for the birthing process. If you are a Christian, it is an uncomfortable fact having God responsible for the poor design of women, which leads to the aborted deaths of over a hundred thousand babies each year. With God being responsible, he would be a monster abortionist.

There are thousands other cruelties witnessed in the natural world every day, but there was one activity of an insect that particularly bothered Darwin, which he revealed[24] in a letter to a friend:

I cannot see, as plainly as others do, evidence of design or benefi-cence on all sides of us. There seems to be too much misery in the world. I cannot persuade myself that a beneficent and omnipotent God would have designedly created the Ichneumonidae {a wasp} with the expressed intention of their feeding within the living bodies of caterpillars, or that a cat should play with mice.

Darwin could argue that God's plan is not the total answer to our morality. We should also use our experiences in addition to our evolved moral sense.

CHURCH RESPONSE

By the mid-1700s, bedrock Christian theology started addressing the emerging natural sciences that were indicating that the Earth was much older than given in the biblical text. Later in the century, William Paley (1743-1805), an English clergyman, stated his belief on man's creation and argued that man looked as if he was designed, and only his supernatural Christian God could have been the designer. This argument for an intelligent designer was to become a strong Christian position; as Paley said, defending God as the designer:

There cannot be design without a designer . . .

Paley's teleological argument was based on observing humans and all of their complexity, who were to him obviously designed by his Christian God. This was the prevailing Christian belief that Darwin encountered in the mid-1850s.

After Darwin's book was published, Bishop Samuel Wilberforce, Bishop of Oxford, lectured in 1860 against Darwin's "Atheistic theory" based on Paley's "designer" arguments. Arguing on Darwin's side was Thomas Huxley and Joseph Hooker. Huxley's first lecture proved to be just the beginning of a debate that continues to this day. Huxley, known as "Darwin's bulldog" in the debates, advocated for an "Ape Origin of Man" and against the church's argument for a "God Origin of Man." Paley established the mindset that God was the designer of man. Paley did not attempt to separate the natural from the supernatural worlds and continued with God being in charge of both.

Paley's design argument made it to America, where in 1874 before Darwin's death, a theologian, Charles Hodge, published *What is Darwinism?*, which attacked Darwin's Theory as being thoroughly materialistic and Atheistic by claiming that:

The denial of design in Nature is actually the denial of God.

Paley's argument for design has been used by many Christians since then. Since Darwin, Christian debaters have hurled words such as "Atheists" and "ape man" repeatedly at supporters of Darwin's Theory out of fear of losing God's authority over science. But slowly, over the last half of the nineteenth century, Darwinian evolution became the standard accepted by the scientific community.

This position was challenged publicly after the First World War with the rise in America of Christian Fundamentalists, who worked to have state laws passed banning the teaching of Darwin's Theory in public schools, the Scopes trial in 1925 being the most famous. At first Christians fought to keep man's origins and design described in Genesis as the one to be taught in public schools. This approach is known as Creationism. Attempts to get the Creationism theotheory into the public education arena resulted in court cases, which ruled against their acceptance and in favor of keeping religion out of public education. Since that time other theotheories that mix in a little science have been introduced, and they have also failed as a result of being bad science, bad theology, and a violation of constitutional law.[62]

The failures by Fundamentalists in public schools and federal courthouses to sell their theotheories have obscured the progress made among Liberal and Progressive Christians. They moved away from applying a literal reading of Genesis toward to a more metaphorical position. However, the Liberal Christian position on science remains God-centric while Progressive Christians have taken the step to separate their God's supernatural world from Nature's world. One group of Christians, The Clergy Letter Project, has defined their position as respecting both religion and science:

We believe that the theory of evolution is a foundational scientific truth, one that has stood up to rigorous scrutiny and upon which much of human knowledge and achievement rests. To reject this truth or to treat it as "one theory among others" is to deliberately embrace scientific ignorance and transmit such ignorance to our children . . . We urge school board members to preserve the integrity of the science curriculum by affirming the teaching of the theory of evolution as a core component of human knowledge. We ask that science remain science and that religion remain religion, two very different, but complementary, forms of truth.

There is continued opposition from the more fundamental Christians as they continue to resist any acceptance of Darwin's Theory and strongly advocate for God's authority over Nature. Examples of such positions are found in Fundamental Christian colleges[63] today, whose position on natural science is that:

True science will fit that (biblical) framework: anything that fails to fit the biblical framework must be rejected as erroneous.

Censorship of science by religious authority cannot be expressed in stronger terms, but such Fundamentalists are a decreasing minority. Many Christians are more tolerant of secular science that "fails to fit the biblical framework," as they are beginning to change the framework to reflect an increase in metaphorical acceptance of parts of the Bible. Today the contrast between the Lutheran positions and that of a Fundamentalist Christian college illustrates the progress being made by some Christians to accept secular science and separately their religion.

To preserve their own integrity both science and religion need to remain in a healthful tension of respect toward one another and

to engage in a searching debate which no more permits theologians
to pose as scientists than it permits scientists to pose as theologians.

This liberal position is reflected in five of the Christian books reviewed, but they fail to give full acceptance to Darwin's Theory—that is, they continue to resist removing God's authority over science. Liberal theotheories on evolution remain God-centric, a view that reflects the general view of the population. A 2010 Gallup poll found that 40 percent of Americans believe in Creationism, 38 percent in Theistic Evolution, and only 16 percent in Darwinian evolution.

THEOTHEORIES

Confronting Darwin's Theory in the early twentieth century, Christians were confident that they could sweep it aside, or as William Jennings Bryan's said, "drive it out of the country." In one notable case, the state of Tennessee had voted in a law that prohibited the teaching of Darwin's Theory in public schools. This was right up Bryan's alley, and in a short period, Bryan was able to mount a national campaign and have seven states pass comparable laws making it illegal to teach Darwin in public schools. In the first test case, John Scopes, a schoolteacher, was brought to trial in 1925 in a small town, Dayton, Tennessee.[64] By criminalizing the discussions of Darwin in science classes, Bryan thought he could achieve his goal of running godless science out of public schools and out of the country. But the Scopes trial turned into a national case and showcased the confrontation of two famous men arguing an interesting debate and it received national visibility. It proved to be high drama at the national level. For science, there was the famous defense lawyer, Clarence Darrow, defending Scopes. For the Christians and the state, the equally famous politician William Jennings Bryan, a former candidate for president, took the stand and defended biblical creation. Bryan argued like an old-time preacher trying to protect Americans from those who "have no other purpose than ridiculing every Christian who believes in the Bible." Although Scopes was found guilty and fined $100, he immediately appealed, but the case died[72] in the appeal process, for Christians did not want to continue the national attention on evolution by the combative Darrow.

Christians had won their case and the Tennessee law was upheld, but national publicity gave natural science an increased favorable rating. The trial succeeded in getting the attention of both sides, knowing that there were going to be more battles to fight in the future; and indeed, that is what has happened. Christians took the approach that they could replace Darwin's Theory with invented theotheories describing the creation-evolution of man with their God as the creator and designer. With a veneer of science on a new theotheory known as Scientific Creationism, the Christians returned and attempted to get it taught in public schools.[65] They failed.

Later, other Christian theotheories, such as the well-publicized Intelligent Design theotheory, were put forward, and they have also failed to win over the scientific community and public universities or to prevail in the courts. Many Christians who are scientists have also rejected them. Yet some Christians, including five of the authors reviewed continue to push their God-centric theotheories in the public domain. For example, one of the authors, Francis Collins, in his book, and others, including many Catholics, are proposing the Theistic Evolution theotheory.

The continuing attempts by Christians to push newer supernatural theotheories in public institutions are suffering the same fate as the previous ones. To date, the Theistic Evolution theotheory has been discussed in books and articles by several Christian authors, but it has not been pushed for use in secular public schools, so it has yet to be formally challenged in the courts, where it is expected to fail.

Some of the theotheories proposed as Christian answers to Darwin's Theory for use in the public sphere, including public schools, are summarized below.

CREATIONISM

The Creationism theotheory is taken directly from the Genesis, a description of God's creation of man, Adam and Eve, and his design of man in God's image. This theotheory clearly violates the separation of the church and state concept for use in public schools. The first public trial of the Creationism theotheory was at the Scopes trial. Subsequent cases that attempted to inject Creationism into public education as a Christian answer to Darwin's theory have failed in public schools, universities, and courtrooms.

SCIENTIFIC CREATIONISM

After the success of the Russian Sputnik satellite, Americans rushed to beef up their science textbooks. The biology textbooks were found to be in need of updating, particularly their handling of evolution. The National Science Foundation, which funded the Biological Sciences Curriculum Study, corrected this with new textbooks in 1963 and included the subjects of evolution and sex.

Creationists, particularly Henry Morris, responded by proposing an alternative view of Creationism by adding a little science, renamed it Scientific Creationism and put this revised theotheory forward for use in schools.

For decades after the Scopes trial, Creationists had been quiet on the national scene. In 1961, the book *The Genesis Flood: The Biblical Record and Its Scientific Implications,* by Henry Morris and John Whitcomb Jr., produced a spark that renewed the movement to have man's creation-evolution described by Christians. The Scientific

Creationism movement grew in strength and the revised theotheory increased the hopes that the arguments in the book would gather some scientific research. But the expected impact on science fizzled out. Karl Giberson noted:

> *Not a single scientific paper has been published on Scientific Creationism explicitly promoting any aspect of this brand of Creationism in a scientific journal. The movement faltered and in 1987 a Supreme Court ruling[66] refused to grant equal time to religious approaches in the public schools.*

These confrontations resulted in some Christians trying to take a step beyond the Christian Fundamentalists' Creationism theotheory by attempting to add a little more science to later theotheories. A quote from the BioLogos site describes the partial transition:

> *BioLogos exists primarily to help those same young people realize that their Christian faith need not be tied to Mr. Ham's {Fundamentalist} view of a 10,000 year-old earth. God's Word and our life in Christ are much more profound than that! If Mr. Ham and his followers want to continue to believe in a young earth, that is up to them. However, implying that one's life in Christ depends upon holding a particular view of the earth's age is not fair to our young people.*

This statement is a modest but positive step away from Creationism and the literal reading of the scriptures on man's creation and a partial step toward the Darwinian view that the evolution of the species, including man, occurred over an extended period—billions of years.

The dilemma for Liberal Christians is that in order to consider evolution, they must inject their God into Darwin's Theory and reject the literal reading of creation in Genesis. In effect, they must invent a Christian theotheory of evolution that has God directing evolution. Their lack of scientific constraints makes inventing a theotheory an easy task, but reconciling it with science and the scriptures is not. By contrast, Darwin's Theory is constrained by Nature's laws and remains the same when viewed by any person.

When Creationism and Scientific Creationism failed in the courts for being clearly Christian biblical theotheories, other theotheories were subsequently proposed.

INTELLIGENT DESIGN

The next theotheory aimed at replacing Darwin's Theory was the argument that the designs of living creatures are far too complicated for Nature using natural selection to have produced them, and they could only have been made by an Intelligent Designer (ID). This is a reintroduction of Paley's designer arguments of 1802. This approach was picked up by a law professor, Phillip Johnson, who through his book *Darwin on Trial* (1991) took a strong anti-natural selection approach, one that appealed to many Christians, to sack Darwinism. Johnson's book excited many Christians and he was able to receive financial support from the Discovery Institute for selling the theotheory. Additionally, he received general support from several Christians with science degrees who stepped in to work on the new Intelligent Design theotheory. One of those, William Dembski, joined the effort and described the argument for ID:

> *There exist natural systems that cannot be adequately explained in terms of undirected natural causes and that exhibit features which in any other circumstance we would attribute to intelligence.*

Dembski's use of the phrase "undirected natural causes" and features attributable to intelligence indicates that he fails to understand that Nature designs without a designer. The central ID argument was that many biological features, such as the human blood clotting system, the bacterial flagellum, and the 600 volt producing eels are irreducibly complex—that is, too complex to be produced by natural selection.

Nature can indeed be complex, but as scientists have demonstrated, it is reducible to steps which can be accomplished by natural selection. The evolutionary biologist Russell Doolittle dismisses the arguments that there are "irreducibly complex" living bodies in Nature. Doolittle and others have researched the evolutionary steps that have produced the complicated blood-clotting mechanisms from early life (amphioxus) forms to humans. For humans it is a complex system involving more than two dozen extracellular proteins that must be switched on and off at the right time to make the clotting process work. Nature is indeed a master designer. An overview of this work is outlined in Appendix D: Overview of the Evolution of Vertebrate Blood Clotting, taken from his book by the same name. Judge Jones, presiding in the Kitzmiller case in Dover, Pennsylvania, case on using ID in public classrooms, wrote in his trial summary:

> We therefore find that Professor Behe's claim for "irreducible complexity" has been refuted in peer review research papers and has been rejected by the scientific community at large.

Thus, the attempt by Christians on a local school board to use ID in public schools suffered a major defeat in this widely publicized case. The ID theotheory suffers a fate comparable to that given to earlier theotheories by the scientific community and by the courts as the Dover case has demonstrated.

SACRED INNOVATIVE DESIGN
The decisive defeat of the Intelligent Design theotheory did not reduce the efforts of Christians to propose other theotheories.

Michael Dowd proposes a comparable ID theotheory, which I have given the name "Sacred Innovative Design", as his replacement for Darwin's Theory. It is structured like the previous theotheories by mixing religion and science theories together and declaring that all science, including evolution, is sacred. By this declaration Dowd gives God the authority over all science, including Darwinian evolution, thereby allowing him to remove chance used in Darwin's Theory with God's certainty. He declares that he has converted natural selection to "intelligent innovation" but in reality he is only continuing with the intelligent design argument rejected by other Christian scientists-authors, such as Collins, Falk, Giberson, and Ayala.

THEISTIC EVOLUTION

The Theistic Evolution theotheory proposed by Francis Collins and others, including some members of the Catholic Church, makes evolution God-centric. Although it varies in some details between believers, the Theistic Evolution theotheory uses some tenets of Darwin's Theory and rejects others. Collins uses the argument that God is the authority over man's evolution, and it is part of God's plan:

> God's plan included the process of evolution and natural selection permitted the development of biological diversity and complexity over very long periods of time.

In effect, Collins' argument is but an attempt to hijack Darwin's Theory by calling it "God's plan." He accepts some of the theory but installs his supernatural God as the creator and designer who uses miracles here and there. The Catholic version of Theistic Evolution has God inserting a "soul" into man. They give no scientific support for the insertion, so they use a miracle to impose an unexplainable supernatural act on Darwin's Theory.

Summary

Attempts by Christians to replace Darwin's Theory with religious theotheories on evolution have failed in the classroom for being unconstitutional and have failed to get the support of the scientific community for being bad science. If we paraphrase Senator Moynihan's comment on opinions and facts in terms of God's supernatural world of theotheories, we have:

Everyone can have their own religious theotheory, but not everyone can have their own natural science theory which requires approval by the scientific community.

In the books reviewed three theotheories are employed: Sacred Innovative Design (Dowd), Theistic Evolution (Collins), and Intelligent Design (D'Souza). However, these theotheories seem like old movies we have seen before in the conflicts of religion with science. The authors as dutiful Christians venture forth to fight the same battle with the forces of science (evil?) again using the same old approach—place God in charge of the science process (natural selection) and mix science and religion to develop a new theotheory. As before they are rejected by the scientific community for not accepting Darwin's Theory in full. Table B gives a comparison of the acceptance of six tenets of Darwin's Theory with four theotheories.

However, the lack of acceptance of Darwin's Theory ensuring the rejection of the theotheories by the scientific community does not seem to dampen the urge and passion to continue to make theotheories on evolution God-centric, rather Nature-centric as Darwin did.

Table B—Theotheories and Acceptance of Darwinian Tenets					
	Creat ionism	Intell Design	Sacred Innov	Theistic Evol	Darwin Evol
1. Natural process	no	no	no	no	yes
2. Long time frame	no	yes	yes	yes	yes
3. Common ancestor	no	no	yes	yes	yes
4. Design by Nature	no	no	no	no	yes
5. Random mutations	no	no	no	no	yes
6. Chance outcomes	no	no	no	no	yes

Instead of battling, Christians and their churches have a vast potential for expanding much needed humanitarian services in addition to their religious services. This is, of course, a central message given by Jesus. It does not seem that arguing over the details of Darwinian evolution is supportive to Jesus' message. God and Nature both have roles to play. I believe St. Augustine would quickly step into this argument and add that God's role is to direct believers to salvation and to providing humanitarian activities. Christians should stop wasting their time arguing theotheories and let science explain scientific processes, such as, evolution.

Neuronian Revolution

The two earlier scientific revolutions, Copernican and Darwinian, were watershed events advancing scientific knowledge of Nature and producing new scientific theories that replaced existing church dogma in several areas of science. From the Copernican Revolution came a physics theory, the helio-centric theory, which replaced the Church's less accurate geo-centric theotheory. In the second, the Darwinian Revolution, Darwin's Theory of natural selection replaces the Christian supernatural theotheory of man's creation and design.

A third scientific revolution, the Neuronian Revolution, is currently underway, led by neuroscientists who are providing a Nature-based explanation for the evolution of the capabilities of man's brain (a matrix of neurons) to produce his tools, his morality, his spirituality, and his mental ability to invent supernatural gods. By the end of the twentieth century, neuroscience was beginning to make inroads to understand how the biological brain could provide explanations of the evolution of man's morals and spirituality. Patricia Churchland[29] outlines a naturalistic perspective for the evolution of morals in her book, *Braintrust*. She notes that it can:

> . . . *help us disentangle ourselves from many myths about morality. In disentangling ourselves from the myths, we may become even more keenly aware of our obligation to think a problem through rather than just react blindly or follow a rule.*

A naturalistic perspective of the brain is based on research on the neuron and the neuronal network's structure of billions of interconnected neurons. The brain produces a mind that exhibits consciousness, memory, reasoning, emotions, social behavior, moral behavior, and religious belief. Since the mind is the receiver, integrator, translator, storage bin, inventor, and transmitter of all of our information and experiences. Francis Crick postulated in his book *The Astonishing Hypothesis: The Scientific Search for the Soul* that it is our mind that gives us consciousness, morals, and our unique identity—in essence, our "soul." Crick's hypothesis:

> *You, your joys and sorrows, your memories and ambitions, your sense of personal identity and free will, are in fact no more than the behavior of a vast assembly of nerve cells and their associated molecules. As Lewis Carroll's Alice might have phrased it: "You're nothing but a pack of neurons."*

Biological evolution gives us our "pack of neurons," the network of billions of neurons, in our brain, the largest of all animals, which gives us our thoughts. Neuroscience describes how these neurons function in networks, the extent of their inherent capabilities, and their flexibility to be enhanced by education and experience. Explanations of consciousness are proving elusive, and more work is needed to produce a fruitful cognitive and emotional framework. But progress is being made on many fronts—the understanding of the range of man's evolved mental capabilities, including his moral sense.

Philosopher Paul Churchland has summarized the functional understanding of neurons, neural networks, and connections to the senses in his book *The Engine of Reason and the Seat of the Soul: A Philosophical Journey into the Brain*. Churchland explains how moral knowledge, like science knowledge, is built from the bottom up. Neural networks are formed that can be rewired and tuned to meet new conditions. This allows man to learn from his moral mistakes and correct them. It is the same approach successfully used by the scientific community to build up scientific information a little at the time with continuing corrections. Moral knowledge is then:

. . . real knowledge precisely because it results from the contin-
ual readjustment of our convictions and practices in light of our
unfolding experiences of the real world.

The Neuronian Revolution is providing information to expand our understanding of how man's brain helped his survival.[28] Evolution has given us a broad set of mental capabilities for cognition and social interactions, which are used in our moral development. Advances in neurobiology are increasing our knowledge of a broad range of human activities, from personal moral responsibilities to the moral actions of religions and governments. Understanding the evolution of spiritual-ity and morality is essential to answering the question we posed: *Who to thank for evolution—God or Nature?*

The functioning of the brain is vastly complicated—a collection of many smaller "brains" interconnected into larger neural networks, which produce various mental states, some conscious and some unconscious. It is known that evolution has left an imprint from its ancestors on our brain, for some of present day functions can be traced to earlier stages in evolution, such as the reptilian brain of our early ancestors.

Although the decision-making process in the brain is not totally known, it is the mind that makes decisions, including moral ones, based on many things (stored information, emotional context, the physical and chemical status of the brain, etc.). How man makes decisions can, in turn, impact man's morality, social behavior, and religiosity. In time of stress, older regions of the brain, such as the reptilian, may be called into action when fast response times are needed. How each of these regions functions, how each is intercon-nected (by chemicals as well as electrical impulses), and how they are managed by higher level executive regions of the brain are subjects of ongoing research.

MORALS

Christian morals presented in the scriptures are and have been greatly admired by many. Francis Collins, one of the authors whose book is reviewed, was attracted to Christianity by the moral laws in the scriptures and said that these were a major factor in his conversion to Christianity. Many Christians consider the morals in the scriptures, the Ten Commandments, Jesus' parables, and others to be the base for Western morals. The importance and value of Christian morals to hundreds of millions of believers is to be respected. The morals described in the scriptures cover a wide range of guidance to believers. There are a number of morals and rules in the scriptures, such as the Golden Rule and the Ten Commandments that are widely used by many religions, some preceding their use in the scriptures. Some have been adapted from other religions and cultures, indicating their general usefulness to all peoples. An example of the utility of the Golden Rule for Christians was explained by a retired Los Angeles minister who had administrated to the poor on the streets of Los Angeles for thirty years. When asked by a reporter about his experiences, he summarized his ministry by saying that the success he enjoyed was based on getting the Golden Rule of Jesus to the people. The fact that he could impact the lives of many through communicating this simple message underscores the importance of these general morals. Finding moral messages that are demonstratively helpful to others highlights the skill of the minister and the value of the moral being taught.

The Ten Commandments are a version of social morals, which had been used by surrounding societies for many years. For example, in Babylonia a thousand years earlier, around 1780 BCE, King Hammurabi had inscribed on stone steles 292 laws for overall guidance to the people of his empire. Hammurabi is reported to have received these laws from his god, Shamash. Other kings have had comparable moral commandments from their gods.

MORAL EVOLUTION

Human morality is the core of being human, a declaration on which both Christians and Naturalists would agree. It is real knowledge that is used in decision making on a daily basis. Through feedback, man's morals continue to evolve. For two thousand years, the Christian church has preached that man's morals were given to him by God. The Neuronian Revolution is offering an alternative source: Nature—that is, Darwinian evolution is the mechanism by which our morals have been developed. This book is not the place to discuss the many thorny ethical issues facing man and our understanding of them, only that it is important to recognize that human evolution has given us a biological basis for our ethics. Thus, two sources of morals are available for us to consider in our discussions: First, Christians have God's gift of moral laws from a supernatural God on high down to man. Second, Nature's evolution of the morals is from the bottom up from evolution and experience. Both can be modified by man's interactions and experiences in the natural world.

In Nature's world morals are designed without a designer, for they are a product of Nature's processes. When moral mistakes are made through trial and error, they can be changed, and thus, over time, mankind's broad range of moral experience can be used. This process offers an advantage over gifts of morals from God that cannot be corrected, for who can tell God to update his morals? This has been particularly obvious, with many of the biblical social morals becoming conflicted with cultural changes over time.[30] The recognition of Nature as a source of morals does not negate the importance of God's morals for Christians who have two sets of morals for their use.

Darwin looked at morals from a broad evolutionary view and argued that humans and nonhumans share much in common. In *The Descent of Man*, he argued that there is continuity in the evolution of morals over the range from primates to humans and speculated that the fundamental bases for morals are the social interactions.

> *The so-called moral sense is aboriginally derived from the social instincts. Social instincts lead an animal to take pleasure in the society of its fellows, to feel a certain amount of sympathy with them, and to perform various services to them.*

Much has been learned since Darwin's time. At the gene level, we have learned that genes can be selfish and deceitful in their quest for survival, but we have also learned that survivability may be enhanced by man's morals derived from group social interactions. The evolution of morality involves many factors, which in certain combinations may not lead to the highest moral behavior for individuals, for compromises are required in the development of group or institutional morals, such as our democracy.

Nicholas Wade has summarized the foundation for moral evolution:

> *Among primates, including humans, social behavior has a substantial genetic basis. Research suggests that sociality emerged about 52 million years ago. The earliest primates sought safety by being solitary and inconspicuous, moving only at night. It seems that when they shifted to daytime activity, they sought safety in numbers. It was from these loose, unstructured groups that more specific forms of primate social behavior began to evolve, some 16 million years ago. These included pair bonding, an arrangement adopted by gorillas and humans.*[31]

The impulse to help others may become habitual, and the community may play a key role in forming these habits. In fact, says Darwin, the social instincts, along with the aid of the intellect and social impulses, lead naturally to the Golden Rule ("Do to others as you would have them do to you"), a widely used moral. The evolved moral sense, which includes altruism[26] and cooperation[27] is observed in our primate ancestors as well as man. Patricia Churchland argues that from our "pack of neurons" evolution has given man:

> ... *morality {that} is grounded in our biology, in our compassion and our ability to learn and figure things out.*

Now, some 150 years after Darwin opened the door for understanding the natural evolution of morals, Churchland summarizes her understanding of morality today:

> *Morality seems to me to be a natural phenomenon—constrained by the forces of natural selection, rooted in neurobiology, shaped by local ecology, and modified by cultural developments.*

Building on the biology of the brain and its structure, Churchland outlines the natural evolution of morals:

> *Moral values are rooted in a behavior common to all animals, the caring for offsprings. The evolved structure, processes, and chemistry of the brain incline humans to strive not only for self-preservation, but for the well-being of allied selves, first offsprings, then mates, kin, and so on, in wider and sider "caring" circles. Separation and exclusion cause pain, and the company of live ones causes pleasure: responding to feelings of social pain and pleasure, brains adjust their circuitry to local customs. In this way, care is apportioned, conscious molded and moral intuitions instilled.*

These findings are supported by a broad range of experiments, from animals to man, exploring the biological base of morality. As an example Churchland describes experiments with a small animal, the vole, that illustrate the behavioral effects of certain chemicals on the brain, in this example, oxytocin and vasopressin. Tests showed that they have a pronounced effect on the sociability of two different vole species. Prairie voles mate for life, and montane voles do not. Among prairie voles, the males not only share parenting duties, they will even lick and nurture pups that aren't their own. By contrast, male montane voles do not actively parent even their own offspring.

What accounts for this difference? Researchers have found that the prairie voles, the sociable ones, have greater numbers of oxytocin receptors in certain regions of the brain. Prairie voles that have had their oxytocin receptors blocked will not pair-bond. In dogs and humans, it has been found the level of oxytocin in the brains of both increases when man and dog are socializing, or bonding. Churchland notes in man that oxytocin acts:

> . . . *by decreasing the stress response, allows humans to develop trust in one another necessary for the development of close-knit ties, social institutions, and morality.*

In humans, cultural variations of morals tend to hide the social unity to humankind. Oxytocin has become known as the "love hormone," for it and its cousin compounds support the human capacity for empathy. It serves as a trigger for social sentiments; when the personal association is positive, it bolsters pro-social behavior and the opposite when the association is negative. The links of oxytocin and other chemicals to the activation of the reward system of the brain are complicated and are now being studied in detail.

Oxytocin is an ancient chemical[32] found in reptiles, birds, and mammals that has long evolved with its hosts for over hundreds of millions of years. It has served many functions in living organisms during this time, including the basic cell function of water balance.

Today it not only impacts the social effects of the vole and man, it is also used today to induce childbirth in humans.

In addition to the evolution of chemicals impacting the brain, the neuron-matrix structure of the brain has also evolved to meet the demands of survival. An example is the brain's adaption to interactions with friend and foe, with the evolution of mirror neutrons in mammalian brains, from monkeys to man and in other species, including birds. These neurons may be an important element of social cognition, for they are brain cells that seem specialized to code the actions of others as well as ourselves. What are the gestures and facial expressions of others really telling us? The more we know about what others are thinking and the faster we know it, the higher our survival rate, for without mirror neurons, we would likely be less attuned to the intentions, emotions, and actions taken by other animals, including people, around us. Marco Iacoboni notes:

> *The way mirror neurons likely let us understand others is by providing some kind of inner imitation of the actions of other people, which in turn leads us to "simulate" the intentions and emotions associated with those actions.*[33]

We know that many animals are capable of detecting the intentions of other animals to one degree or another. Do I have to run and hide from the other animals I see, or can I wait? Mirror neurons appear to go a long way back in our evolutionary tree and have helped our ancestors survive. The input from mirror neurons does not give the full understanding of how the intentions of others are understood, for other parts of the brain are also involved. But it does appear that they are involved in the evolution of sociability, a necessary base for morals.

Dogs, domesticated by their interactions with man, show an increased capability to understand man's intentions and actions, even responding to directions by man's eye movements. Wolves, the immediate ancestors of dogs, on the other hand, do not have the capability for socializing with man, for it has not been possible to domesticate them to date. They remain focused predators, and whenever man shows them any affection, they respond by keeping man in their mind as just one more animal in the food chain. Understanding biological changes in the brain for socializing in dogs can help our understanding of man's moral evolution.

For man and his latter-day ancestors, close cooperation and socialization among kin and tribe increased the survivability of the group. The philosopher Michael Ruse summarized the importance of our evolved moral sense:

> *Our moral sense, our altruistic nature, is an adaptation—a feature helping us in the struggle for existence and reproduction—no less than hands and eyes, teeth and feet. It is a cost-effective way of getting us to cooperate, which avoids both the pitfalls of blind action and the expense of a super brain of pure rationality. .*

In addition to neuroscientists and philosophers, the research of primatologists, physiologists, and psychologists is expanding our understanding of the biological base of morals. Frans de Waal, a primatologist, sees the natural evolution of morality in our ancestors:

> *Morality is often considered as opposite of human nature: our main tool to keep human nature in check. Yet the moral sense likely evolved along with the rest of human sociality.*

From his observation of chimps and other animals, de Waal notes:

> *I've argued that many of what philosophers call moral sentiments can be seen in other species. In chimpanzees and other animals, you see examples of sympathy, empathy, reciprocity, a willingness*

to follow social rules. Dogs are a good example of a species that have and obey social rules; that's why we like them so much, even though they're large carnivores.

In effect, de Waal is telling us that the evolution of morals has proceeded without a sharp boundary between the morals sentiments of primates and those of humans, but along a gently sloping transition:

We start out postulating sharp boundaries, such as between humans and apes . . . {but} evolutionary theory always leads us: to a gently sloping beach.

Further, de Waal notes:

If a human moral instinct developed from these building blocks, then morality has a genetic basis and may well have evolved over the millennia into forms that are objectively higher.

In summary, research with animals by de Waal shows that animals naturally have pro-social tendencies for reciprocity, fairness, and empathy. Evolution extended these tendencies to humans to form a base for morality in tribes and small communities. Understanding expanding morality from small to larger groups has been and remains challenging.

Richard Joyce, philosopher, addresses the evolutionary base of moral behavior as a high-level behavior generated by the brain. This thesis is being tested by applying magnetic fields to a specific brain region, and indeed the results show changes in people's moral judgments with varying magnetic fields, a strong indication that moral judgments are a brain process.[34] As Joyce[35] notes:

. . . our biological evolution of morals is a process that has made us sociable, able to enter into cooperative exchanges, capable of

love, empathy, altruism —granting us the capacity to take direct
interest in the welfare of others with no thought of rectification—
and has designed us to think of our relations with one another in
moral terms.

But evolution has also brought other actions by individuals that encompass cheating, deception, and selfishness, which, in a religious term, is evil. These were combated by group punishment, usually banishment, which without the protection of the group normally led to death. Thus, life through natural selection has evolved balancing bad morals against good morals where usually the good ones prevail, but not always.

Scientists see our morality beginning in our evolutionary past, with our ancestors, rats, monkeys, apes, chimps, and on to *man.* By the time humans came around, evolution had forged a moral sense of right and wrong. This early moral sense has been observed in experiments on human babies[36] as young as six months. Tests on children are reported by Ernst Fehr of the University of Zurich show that:

by the age of 6 or 7, children are zealously devoted to the equitable
partitioning of goods, and they will choose to punish those who try
to grab more than their arithmetically proper share of jelly beans
even when that means the punishers must sacrifice their own por-
tion of treat.

Homo sapiens are but the latest version of moral primates who satisfied our obligations to others in the tribe and through cooperation with others increased their survivability. Natalie Angier, a science writer, summarizes man's moral sense[37] extending into bands of men:

Homo sapiens have an innate distaste for hierarchical extremes,
the legacy of our long nomadic prehistory as tightly knit bands

living by veldt-ready team-building rules: the belief in fairness and reciprocity, a capacity for empathy and impulse control, and a willingness to work cooperatively in ways that even our smartest primate kin cannot match.

Successful socialization was critical for our human evolution and involved thoughts about the interactions of one's behavior with that of others around him. As man's capability for cooperation increased, it allowed for larger and more complex organizations to form.

With moral evolution as part of man's evolution over many millions of years, Frans de Waal has difficulty with the concept of morals as a gift to man at one point in time in the past seven thousand years.

Does anyone truly believe that our ancestors lacked social norms before they had religion? Did they never assist others in need, or complain about an unfair deal? Humans must have worried about the functioning of their communities well before the current religions arose, which is only a few thousand years ago.[38] *Religion is but an add-on rather than the wellspring of morality. No one doubts the superiority of our intellect, but we have no basic wants or needs that are not also present in our close relatives. I interact on a daily basis with monkeys and apes, which just like us, strive for power, enjoy sex, want security and affection, kill over territory, and value trust and cooperation. Yes, we use cell phones and fly airplanes, but our psychological make-up remains that of a social primate.*

Patricia Churchland offers the thought that humans have a capability to do much better:

Humans are able to evaluate a law as a bad law or a good law, without appealing to a yet deeper law.

The Neuronian Revolution is helping to remove the dogma that human morality depends on obeying higher laws, that is, laws sent down from God. Understanding the natural evolution of morality has given man the freedom to evaluate laws and rule in light of their evolved moral sense. Does this not give man a sounder base to his morality? It is up to each individual to express his morality by assessing whether a law is a good or bad one and not blindly follow a fixed law.

William Keith[39] also has difficulty with morals being sent down to man by divine mandate:

> *Basically human beings are moral and sociable, despite claims down the ages that Man is inherently fallen. Our morality wasn't handed to us as divine mandate; it developed slowly and sometimes painfully over the course of eons of evolution, and when we became able to transmit our experience to our children and countrymen, it began to grow faster from example (we are, in fact, very good at copying behaviors that work).*

And Owen Flanagan concludes:

> *The reason we both think it {biologically based morals} makes sense is that the other stories—that morality comes from God, or from philosophical intuition—are just implausible.*

In short, neuroscience is providing an experimental base for understanding a moral sense evolved by Nature's bottom-up process, starting with the parental caring of offspring's hundreds of millions of years ago, to kin, to kith, and on to larger and more diverse groups. Morality evolved from family caring to socializing and to ethical interactions. This, coupled to man's search for spiritual guidance, contributed to the invention of gods. The evolutionary origin of human morality and spirituality leads inexorably to the probability of a human origin of god narratives and god figures.

BIBLICAL MORALS

There are a number of morals from God of epic proportions. An example is Noah and the global flood. God sent a global flood to earth for man's disobedience and proceeded to kill all peoples (including innocent babies and children) on the earth except Noah and his immediate family.

Another example is Adam and Eve's disobedience to God's commandments. God banished them from the Garden of Eden, and their "Original Sin" was commanded by God to be passed to all of mankind forever thereafter.

Many biblical scholars argue that epic morals are metaphors written into the scriptures to illustrate the power of God and should not be taken literally. Some Fundamentalists still argue they are to be taken literally.

Some biblical morals reflect the general morality of Middle Eastern societies at the time the Jewish Bible was written and later when the New Testament was written. Some of these morals have been subsequently written into laws in Christian countries. As societies have changed, most of these moral laws are no longer acceptable. Today, some have been rejected by our democracy. However, they remain in the scriptures and, from time to time, cause problems for Christian literalists. Three examples of biblical Christian morals that have been replaced by secular laws are:

> *Accepting slavery, denying women rights (voting and civil laws), and denying homosexuals other rights.*

Bishop Spong, a biblical scholar, notes problems with some biblical morals seen in light of today's society:

The Bible is no longer even regarded as moral. Its verses have been used to affirm war, slavery, segregation and apartheid. It defines women as inferior creatures and suggests that homosexual persons be put to death.

Slavery is an example of a morally acceptable act in the Christian narrative, but one that was rejected by secular governments over a hundred and fifty years ago—by England in 1833 and the United States in 1865. Our Constitution has been amended on the issue of slavery, the Bible has not.

Examples of persons identified as evil, including witches and homosexuals, are unacceptable in our democracy today. In the Middle Ages, the church made it one of their missions to root out heretics by branding them as immoral people—witches—and killing them, usually by burning at the stake. Most of the time, there was no substantial evidence required to identify a person, usually a woman, as a witch. At the height of Christian theocratic power, the search and elimination of witches and other persons of "evil" moral character were carried out on a grand scale. Before the Vatican program phased out in Europe, thousands of witches were accused, tortured, convicted, and killed.

In the US in the late 1600s, church men and women in Salem, Massachusetts, came to believe that many misfortunes were attributed to the work of the devil: infant death, seizures, crop failures, or friction among the congregation. In cases the supernatural was blamed. All of this came together and fueled the craze of 1692, which resulted in two hundred accused and 150 sent to prison, where five died, nineteen were hanged, and one was pressed to death. George Lincoln Burr has noted that it was not long before the people felt remorse and shame for the actions of their theocratic governance directed by their ministers and councilmen.

More than once it has been said, too, that the Salem witchcraft was the rock on which the theocracy shattered.

By 1700, most countries had decided that the church's activities against witches had gone too far and rejected the church's religious authority and the justification of biblical morals to name and prosecute witches. Later after three hundred years, the Salem trails were revisited, and all were finally proclaimed innocent by the state Governor Jane Swift. No wonder there is disagreement within the Christian community on the "truth" or validity of biblical morals.

PARABLES

Some scholars suggest that the moral jewels of the scriptures are those given by Jesus through his parables; they identify the essence of Christianity. They are venerated not only by Christians but by many others. For example, these morals exhort humanity, such as "love thy neighbor" and remain as timely and relevant today as when they were first recorded two thousand years ago. Jesus taught basic moral principles rather than adherence to strict interpretations of the Jewish law; for example:

> *You have knowledge that it was said, Have love for your neighbor, and hate for him who is against you: But I say to you, Have love for those who are against you, and make prayer for those who are cruel to you, (Matt. 5:43-44)*

Jesus' parables are a continuing source of inspiration for many. They were singled out for personal notice by Thomas Jefferson, a Deist. Jefferson thought about many things, not only founding a new country, but the enhancement of his personal philosophy. His library was the largest in the country at the time and contained many books by philosophers and religious writers and included the Bible and the Koran. From his study of his Bible, Jefferson found much that he wished to reject, such as the supernatural and the epic and societal morals, but he also found the morals of Jesus to be "gold" that he greatly admired. As part of this personal philosophical quest, he edited the Bible and separated the parts to retain that were helpful to his philosophy by cutting out[67] (literally with a pair of scissors) and saving those passages which he thought to be the words of Jesus and pasting them in a blank book, which he titled *The Life and Morals of Jesus of Nazareth, Extracted Textually from the Gospels in Greek, Latin, French and English*. In this little book constructed for his personal use, he rejected references to the supernatural but saved what he thought were Jesus' morals useful to humanity. Jefferson noted:

> *I separate, therefore, the gold from the dross; restore to Him the former, and leave the latter to the stupidity of some and the roguery of others of His disciples.*

To Jefferson, morality arose not from revelation or inspiration, but rather from the dictates of Nature and reason. His editing the Bible for expanding his moral philosophy is an example of using reason, not religious servitude, to arrive at those morals (the gold) of importance to him. We may not be as thorough as Jefferson in developing our moral philosophy, but nonetheless, each person does the same figuratively and develops his own philosophy, some of it based on the work of others, including religions.

It is interesting to note that Jefferson arrived at a similar view, albeit from almost an opposite starting point, a Deist instead of a Catholic, as that of St. Augustine of Hippo, who had expressed a comparable philosophy some 1,300 years before:

> *I have read in Plato and Cicero sayings that are wise and very beautiful; but I have never read in either of them: Come unto me all ye that labor and are heavy laden.*

The approval by St. Augustine and Jefferson (and many others in between) attests to the lasting importance of Jesus' morals.

ORIGINAL SIN

This is a theological concept from God's supernatural scriptures addressing the sinfulness of man. The Genesis narrative has God creating Adam and Eve, warning them of sin and watching frail man and woman fail by disobeying God's commands, thereby committing the "Original Sin." Christians are told not only that Adam and Eve sinned, but that their sin was against all mankind and was to be passed on by mothers, generation to generation, by God's command. Some Christian scholars have questioned the biblical "Original Sin" concept in light of Darwinian evolution. Bishop Spong[68] notes:

> *Darwin had rooted human life in the struggle for survival, which is a mark of all living things. Christians have, however, interpreted this survival drive and its inevitable manifestation of self-centeredness as the mark of our "sinfulness." They understood this mythologically as "The Fall" and this they left human lives victimized by "original sin." On the basis of this analysis of the source of human evil, traditional Christianity has postulated Jesus as the divine rescue operation mounted by the external deity.*

> *This has resulted, I believe, in a theology rooted in victimization with Jesus being seen as the first victim. Catholic Christians refer to their liturgy as the "sacrifice of the mass" and suggest that the mass serves liturgically to make the death of Jesus ever available to overcome the "sin" that is present in us all. Protestant Christians, working in this same theology of victimization, have developed the mantra "Jesus died for our sins."*

Did God give us "Original Sin," or does Nature's evolution result in humans being intrinsically complicated with the ability to be both sinful and good? Christians believe in the first, while naturalists point to humanity's biological evolution over millions of years to provide the base for our evolved sinfulness and goodness. First, our early ancestors competed selfishly among themselves and all others for survival. Later, some became altruistic, and groups of altruists were able to win over selfish individuals, setting the stage for societies to appear. Our genes have the imprint of these evolutionary steps. E. O. Wilson argues that:

> *Within biology itself, the key to the mystery is the force that lifted pre-humans social behavior to the human level. The candidate in my judgment is multilevel selection by which hereditary social behavior improves the competitive ability not of just individuals within groups but among groups as a hold.*

Since man is a complex composite of all of his evolved traits, such as selfish individual survival and altruistic social behavior, how do we determine where in the evolutionary tree of man's ancestors stretching over three billion years an "Original Sin" occurred? It makes no sense to attribute sin to early mammals that did not possess the brainpower to do much more than fight for survival for themselves and protect their kin. Where in the chain of pre-humans before Homo sapiens did cognitive power evolve to have societal rules by which the concept of sin is meaningful?

Nature's view of man's acquisition of morals in the natural world is far different from that of the Christian narrative given their supernatural God. For Naturalists there was no "Original Sin," only an evolution of a moral sense in man's ancestral line from lower animals to primates and on to us. If one views morality as a biological product of man's evolution over billions of years, man's evolved morals can be viewed a "rise to grace" by living organisms. This is a striking difference from the Christian theological concept of an "Original Sin's" "fall from grace."

THE SOUL

The soul is a theological concept useful in the context of the Christian narrative. For Catholics, the soul is immortal and presented as the essence of man to be judged in the resurrection. Their definition is:

The soul is a living being without a body, having reason and free will.

This statement combines the supernatural (a living being without a body) with the natural (human reasoning), and by mixing the two the soul is a theological concept, not a naturally occurring biological one.

When the concept of a soul is presented as the essence of man in physical terms, then science can present Nature's alternative. Neuroscientists Francis Crick and Paul Churchland see our biological mind as the essence of man—that is, his soul. They argue that our essence is found in our brain ("pack of neurons"), a material assembly of cells that have evolved over millions of years through many branches of our ancestral tree. The brain is a material object that obeys Nature's rules (physics, chemistry, and biology). However, for Naturalists, there is no gift of a soul; we have only our evolved brain to call upon. To say that our behavior, or our essence, is based on a vast, interacting assembly of neurons, the brain, should not diminish our view of ourselves, but enlarge it. With his evolving "pack of neurons," his brain, man has not only survived, but has evolved social and moral senses that have produced complex social and ethical cultures and institutions that have greatly improved the lives of man. These achievements are indeed worthy of our deepest praise. Our democracy is but one result of the many steps man has taken to meet the social imperatives from our evolved morals. With flexibility and the ability to learn, our brain offers the capability to adjust to new moral demands of changing societies and meet new challenges in the future.

GODS

Views of gods differ widely when seen through the eyes of Christians or those of Naturalists. Christians believe that their supernatural God (Ex-Nature God) created the universe and man and that man is here only because of the power and grace of the omnipotent God described by the scriptures.

Naturalists see differently; they see the explicable processes of Nature creating living organisms, including man. Through natural evolutionary processes, man's mental capabilities have developed to the point that they produce supernatural religious narratives with gods. Mythical supernatural narratives with gods as causal agents for activities were added by shamans to increase the cohesion to the group by providing a common story that could be shared with other believers. These man invented gods are described as supernatural In-Nature Gods. The view that the supernatural scriptures are a work of human literature was raised by the philosopher Bernard Spinoza in 1640 in his book *Tractatus Theoloico-Politicus*. It was shocking at that time, and to some it is still shocking.

Scientists argue that all concepts of gods are perceived (modeled) by an individual's brain, for man has no other known facility to converse directly with concepts.

Our brains interpret the input from our sensory organs by making a model of the outside world . . . these mental concepts are the only reality we can know. There is no model-independent test for the perception of reality.

We are aware, or believe, only what our minds have modeled, and it is on these models that we base our faiths. Social rules, morals, supernatural gods, religions, and natural science theories are all constructs of models by the human mind. Of all models, only the ones based on science are different, for they can be tested and verified independently of groups or cultures.

Christians have the belief that the input to the brain's model is from their Ex-Nature God by means not known by man. On the other hand, the In-Nature God is an output of the brain's model. Proofs of either one of the Gods are not offered for both are described as supernatural, albeit with vastly different origins.

Christian believers access their God by supernatural belief. Arguments for their supernatural, Ex-Nature God abound in Christian literature and need not be repeated here. Summary arguments for an In-Nature God have only been outlined here as a product of man's evolution.

EX-NATURE GOD

Christian scriptures describe their God as the Ex-Nature God—a supernatural force from out of our universe that created our universe and man. Such creation requires a miracle, for communications from another universe to create our universe before the Big Bang is a task beyond our scientific comprehension.

Once the universe is created, the scriptures have the Ex-Nature God creating man, Adam and Eve, and gifting man with the morals of good and evil. For Christians, the concept of a mysterious, fearful, out-of-the-universe (Ex-Nature) God has continued for two thousand years and continues to be the normative God of Christians today. This is the omnipotent God revealed by the scriptures and modified by subsequent Christian councils and employed by the Christian writers in discussion of evolution in their books.

IN-NATURE GOD

Some Naturalists argue that if Nature is all that there is, then Nature created and designed the universe and life which later evolved into man. Man evolved consciousness, moral and spiritual senses, and mental capabilities sufficient to produce supernatural narratives that included a supernatural God, the In-Nature God.

The morality and humanity included in this God narrative stem from man's evolved social and moral senses. With Homo sapiens there has been a hundred thousand years of human societal experience with kin, tribes, and empires. Jesse Bering[40] notes:

> *We now possess the intellectual tools to observe our own minds at work and to understand how God came to be there. And we alone are poised to ask, "Has our species" unique cognitive evolution duped us into believing in this, the grandest mind of all?*

> *Since the human brain, like any physical organ, is a product of evolution, and since natural selection works without recourse to intelligent forethought, this mental apparatus of ours evolved to think about God quite without need of the latter's consultation, let alone His being real.*

We do know that the human mind has been enriched with human experiences for over a hundred thousand years. Over this period the mind of man has invented many supernatural God narratives that have included an awareness of and insights into the human condition, morals, spiritual aspirations, and an afterlife.[41] Patricia Churchland summarizes the broad base of moral behavior seen in the natural world:

Many people who are not in the least adherents of a deistic religion and may have no theological belief at all are in fact exemplary in their moral behavior. This is also true of whole societies, such as those Asia societies that espouse Confucianism, Taoism or Buddhism, but not a deity that is a law giver.

No one religion or culture has a monopoly on man's morals and religions, which may be with or without gods. History has shown many times over that man is very good at telling stories, including those describing man's interplay with and among supernatural gods. Mythical tales with gods have been used in the works of past storytellers and playwrights, such as Homer, Euripides, Sophocles, and Shakespeare, to name only a few in the West. The Iliad describes a decade-long war between Greece and Troy in the twelfth century BC and was composed and transmitted orally until a written work appeared in the eighth century BC. In the East, the Hindu Vedas[42] in India are early examples of god narratives, dating back to 1,500 BCE.

Judaism became a recognizable religious force about three thousand years ago, with Christianity emerging about two thousand years ago. New religions, such as Mormonism, appeared in the nineteenth century and entered into the competition for believers among established religions. It has been successfully attracting believers with a supernatural narrative different from that of other religions, including Judaism and Christianity, although it uses some elements of both.

When man first started conceiving of gods, he wanted them to be supernatural, strong, and mysterious, for he wished guidance and help from something larger, stronger, and wiser than himself. This need was filled by man's creativity with invented supernatural god narratives. Once gods were invented, shamans used them to gain power by becoming an intermediary between man and the powerful supernatural gods. Man was taught to fear the mighty gods while seeking help from their mysterious power thought to be above what man possessed in the natural world. Robert Wright notes:

> *Religion, having come from the brains of people, is bound to bear the marks of our species, for better and worse.*

There are some people who do not need a god or a supernatural narrative and accept that it is Nature who created the universe and man and all that has flowed from man's creative brain. For these people, Nature is the way to see and experience the natural world and explain the creation of the universe and man. Others turn to their God. It is up to each person to meet his or her spiritual needs through the selection of either God or Nature.

E. O. Wilson sees religion as part of an adaptive design by the evolutionary process, one that comes naturally to man's fertile mind. Further, Wilson notes that there is an ongoing competition for survival between god concepts for believers who have a choice from the many gods available. Further, he also puts religions in competition with scientific naturalism:

> *The final decisive edge enjoyed by scientific naturalism will come from its capacity to explain traditional religion, its chief competition, as a wholly material phenomenon.*

That may be true, but the "edge" Wilson mentions may not be adequate to dislodge the long-believed and deeply held desire by believers for a supernatural god in their lives. So even with an explanation of God as a natural phenomenon (invented by man), that explanation will most likely not be sufficient for many believers to abnegate their beliefs in mysterious and personally supportive supernatural Ex-Nature God. Without definitive proof for either an Ex-Nature or an In-Nature supernatural God the 'true' God is only in the eyes of the believer.

WHENCE EVIL?

With any supernatural god comes the dilemma, whence the source of evil inflicted on man in this world. How a benevolent god can condone evil actions has been an ongoing question by man about his gods. Christians believing in their omnipotent, supernatural, Ex-Nature God know that God must bear the responsibility for all things—the good, the bad, and the evil in the world, such as deadly viruses (HIV, polio, etc.), deadly bacteria (Black Death), malfunctioning human bodies (spontaneous abortion), and deadly physical events (floods, tornados, lightning, etc.), all of which randomly kill and maim millions of innocent people each year.

The polio virus is but one example of an agent that has killed and maimed millions, primarily children, over the years. If God created the polio virus, he must have known what it would do. To combat this deadly virus, secular scientists have devised vaccines and launched global campaigns of inoculations, which have almost eliminated this virus from the Earth. If God is responsible for living organisms and their design, including the polio virus, what are his thoughts knowing that scientists through their research and inoculation operations are in the process of removing one of his creations from the Earth?

This old dilemma continues to be unsettling for an Ex-Nature God. However, the concept of an In-Nature God removes the context of "evil" and "uncaring." One can point to Nature's Darwinian natural selection process as the cause of viruses and bacteria, as well as the poorly designed human body, which produces millions of spontaneous abortions a year. Natural selection does not relate to actions or agents being morally bad, uncaring, or evil, for deaths, healings, and births are just a part of the natural selection process: it is Nature's way of doing its thing.

Christians could adopt this explanation of evil as a natural part of Nature, but the cost would be making Nature independent of God and in charge of man's evolution in the natural world. Are Christians willing to acknowledge that their God is not in charge of the natural world and put Nature in charge of evolution? If not, "whence comes evil?" will continue to remain a dilemma for Christians.

CHURCH RESPONSE

The relative newness of the Neuronian Revolution has not given churches time to take formal positions on the advances of neuroscience. Having the origin of God's supernatural narrative and human morality being the output of man's brain is front and center to answering the question *Who to thank for evolution—God or Nature?*

In essence, this book outlines some of the difficulties that individual Christians and churches have answering the question through literal readings of the scriptures. This difficulty could in the long run be beneficial for both religion and science, for it will lead churches to appreciate the utility of having religion and science separate and independent. Once separate religions are free from the burden of explaining the impacts from advances in natural sciences on the scriptures. The scriptures will be able to stand alone on its own merits and service to humanity.

The State's Way

∞

Religion and the State with its kings, despots, theocracies, parliaments, and democracies, have had a long, interwoven, contentious journey together, filled with continuing struggles over who is in charge of what. For most of the journey, kings and religious leaders have joined into theocracies and exercised authority and power over all matters—government, science, and religion. Ruling power shifted between kings, church leaders, citizen assemblies, and combinations thereof, depending on the era and the personal power of the individual ruler at the time. Early on theocracies pointedly identified and killed heretics of the religious orthodoxy as a means to enhance or retain power: Socrates was killed by orders from the Athenaeum assembly, Giordano Bruno by the Vatican[43] and William Tyndale by the king of England.

Ruling a country requires laws, and in the Middle East, archeological finds of early writings on clay tablets revealed state laws dating back to 4,000 BCE, although laws in verbal form were around much before. But the written form—in this first case cuneiform chiseled on stone tablets—gives us definitive laws in use at the time. In Babylonia, King Hammurabi enshrined on stone steles the state laws, all 292, of his kingdom given to the king by his god. They are in the language of the land and served as the laws for administering justice and assisting commerce throughout the kingdom. For thousands of years, this was the ruling format: the king was anointed by god to rule, and the laws were given by god to the king, who used the laws for governing. There were exceptions, such as the Greeks, who briefly experimented with republics with elected officials, but by and large it was the theocracies with kings that ruled. It took thousands of years for people other than the kings and clergy to have a powerful voice in their government.

THEOCRACIES

Christian theocracies began when Constantine made Christianity part of the Roman Empire with the Edict of Milan in 313, which proclaimed tolerance for all religions allowing Christianity to be accepted throughout the empire. For a thousand years, Christianity gained power with theocracies and became the prevailing form of government in the West. With one religion in power in theocracies, discrimination against other religions was universal. Having morals and laws given by "one's own special God" presented concerns. Churchland notes Aristotle's concerns with theocracies:

> *For one thing, it makes a virtue of intolerance—those who disagree regarding a matter of morals must be dead wrong.*

Moving beyond the intolerance of theocracies took a lot of time and philosophying on ruling by governments. Aristotle is but one philosopher, from others Marcus Aurelius to David Hume, who have declared that divine guidance is not required for morality in government. Theocracies have never been a friend of freedom for the ruling law is by religious dogma and canon law. They come to power by the

force of religion and practice intolerance through the fear of God vengeance on those opposing the theocracy.

Christian theocracies grew and expanded in many Western
countries and were the norm in the West for over a thousand years.
But all large organizations in positions of power, including churches
and governments, are fertile fields for corruption. The Catholic
Church has been no exception, and after a thousand years governing throughout Europe, Catholic theocracies fell into widespread
corruption. An early form of corruption was the selling of indulgences (money paid to get souls out of purgatory)—that is, one
could buy freedom in the afterlife from God's punishment of one's
sins by paying money. The Catholic Church used it widely in the
Middle Ages to extract money from various local states and their
people. Indulgencies were essentially a religious tax by the Vatican
on believers, which they used to build lavish cathedrals, including
St. Peter's and others in Rome.

Opposition to indulgencies became widespread, starting with the
protests by Jan Hus, a Bohemian priest, in 1415. About a century
later, in 1517, a protest by Martin Luther, a German priest and professor of theology, became a central church issue. He had been to
Rome and had seen where the money from the peasants in his part of
Germany was going and wanted to put an end to it. Protests demand-
ing changes in the church went unheeded. These and other demands
for change led many Catholics to a split from the Vatican, causing a
division that culminated in the Protestant Reformation.

The sheer extent and force of the Protestant Reformation from
the 1500s onward caused the Catholic Church to react defensively by
expanding their police force (the Inquisition) to handle internal threats
to Christian dogma and external threats from Protestant governments
in many parts of Europe. Known as the Counter Reformation the

Inquisition grew in size and power as it struggled with the increase in the numbers of believers who were labeled heretics.

At the local level, the spirit of Mythos and Jesus continued as priests serviced the poor and did their humanitarian work, but at the higher levels of the church's organization, there was a secretive and kingly management structure increasingly separating the church bureaucracy from the people. The amount of effort required defending the church from problems of succession, allocation of funds, changes wrought by the Reformation, and internal corruption by the bureaucracy diluted its humanitarian efforts.

To counter these problems, the bureaucracy continued to establish new church dogmas and enforce the old ones. In effect, Christianity had become a big business plagued with all of the systemic human problems of big organizations that we painfully see around us today—poor communications from the workers to the top managers, loss of original goals and bloated compensation for those in charge. Church actions for humanitarian betterment were subjugated to organizational betterment that only produced larger bureaucracies.

During its history there has been and continues to be the human struggle within the Christian Church between those doing good work for the poor and those grabbing power for themselves. The church became more conservative and expended much of its energy upholding the church dogma and a policy of resisting change.[44] Holding on to its power became the dominant mission of church leaders. Those carrying out Jesus' humanitarian message did have their adherents, but they were shortchanged by the upper reaches of the power structure in the church, who mainly directed their efforts to retaining the church's power base.

But all human organizations, however powerful, are changed over time by emerging internal and external forces. In the case of

Christianity, change was driven by the inability of the organization to effectively manage itself. The bureaucracy, who said their actions were by the authority of God, did not welcome feedback from the people on the ground. How can you contest the word of God, you lowly peasant? With uncontestable power and wealth for over a thousand years, the Church bureaucracy became corrupt and disconnected from its believers. Reform from inside proved ineffective, leaving many dissonant groups with the only choice of breaking away from the Vatican and forming their own churches.

It was only to be expected that with the steadily increasing emergence of discoveries in natural science, secular philosophy and governance Christian dogma would come under attack by modernity. With the printing press making information increasingly available to scientists, philosophers, and the rising merchant class across Europe, pressure for change by modernity accelerated. The work by scientists and philosophers gave information with which citizens started exposing, peeling back, and removing some of the many layers of theocratic dogma that had accumulated.

With Christian theocracies losing much of their civil power after the Reformation, state laws were increasingly set by the Common Law (secular) of the People and not by God's laws. Finally, in democracies, the authority for determining bad (evil) persons was removed from religious (God) laws and replaced by secular criminal laws of democracies. "Evil" acts and blasphemy were replaced by specific secular criminal acts against the State, not against the Word of God.

The Catholic Church had never believed that oversight by the hosting nation was proper for the church argued that its morals and laws were superior to those of the host land and best handled by the church. The argument that the church could police itself has dramatically proven to be false and is being challenged in many countries today.

There have been many examples over the years indicating this was the case, but recently the exposure of the worldwide pedophile crimes sadly and dramatically proved that the church was not providing the most vulnerable of its flock, the children, with basic human rights. The crimes against children in the church have been so widespread that countries are pressuring the church to accept government oversight. In many countries, including Ireland a predominately Catholic country, the church's arguments against oversight have been revealed to be indefensible when the light of secular civil governance was shined on them.

DEMOCRATIC REVOLUTION

Philosophers (Aristotle, Marcus Aurelius, Buddha, Confucius, Spinoza, and Hume), looking beyond the ruler of the day, had argued that a supernatural God was not needed for laws, morals, or good government; it was the people being governed who should hold the power, not kings nor churches.

OUR ENGLISH COUSINS

After a thousand years of theocratic rule in the Western world, it was in England that early steps were taken to affect the transfer of power from Christian theocracies to democratic governments. A rebellion by nobles in 1215 forced the king to sign the Magna Carta, a document that limited some of the kings' power over nobles, but not over all peoples. Even so, it has been described as the first of a series of instruments recognized as special to advancing democracy. The others include the Habeas Corpus Act, the Petition of Right, the Bill of Rights, and the Act of Settlement, all of which are major contributions to English Common Law, which Americans inherited. These laws were considered social contracts demanded by the people being governed. They had evolved a little at the time over many steps entailing many struggles by the people to give rights to the people. They were not handed down by God.

The rise of the parliamentary form of government, pushed by efforts to expand the power of the people, was able to continue the transition of power from kings and the clergy to the people. But change was difficult to achieve, for it had to overcome centuries of almost unlimited power by theocracies with king and clergy.

The same difficulties were faced when changes were to be made to religious orthodoxy, such as the introduction of a new translation of the Bible in the vernacular. In England the introduction ended badly for the scholar doing the work. In 1536, William Tyndale, an early scholar and translator, was burned at the stake for his translations of the English Bible after being labeled a heretic. Although he lost his life, his work remains a major contribution to biblical literature, as he was the first to translate directly from the Greek and Hebrew texts. Much of the King James Version of the Bible in widespread use contains his translation. His timely work played a major role in providing biblical material supporting the spread of the Protestant Reformation across Europe.

The empowerment of the people continued in England with the expansion of power by the people's parliament, but it was to be carried to its present democratic level with a written constitution in America.

AMERICA

Five hundred years were to pass after the Magna Carta before a new country in America was formed as a democracy with the People in change. The new nation had no entrenched history of monarchies or theocracies to overcome, only an occupying power, England, to cast off.

Almost every citizen was an immigrant or the son of an immigrant from many different Protestant Christian denominations. This unique population gave the People the opportunity to be the driving force in the formation of a new and democratic government, which was secular and could encompass the diverse population. The Founding Fathers believed that the people (slaves were not included), including immigrants, should be the governing authority.

The Democratic Revolution took its first step with the establishment of our secular Constitution in 1789. Many of the Founding Fathers had experienced theocracies in Europe and hoped that the religious wars and religious persecutions, which had ravished theocracies for far too many years, could be avoided in the new country. It was in America that the Democratic Revolution pulled all of these arguments together and formed a secular government of the people, by the people, and for the people.

In the democracy, the morals and values of the people became an integral part of its laws. Prevailing morals were reflected in the laws formed by the majority of the voters. Democracies defend the right of each individual to exercise his or her own set of morals and values by voting. In aggregate, the laws become secular, for they must serve the people of all religions.

Thus, for a democracy to come into its own and for its citizens to have freedom of religion, religions had to be removed from the government. Churches argued that for governments to be moral, God's laws must be laws of the land. But Enlightenment philosophers outside of the church argued that the people could devise moral governments without God. Additionally, the experience of the English in establishing a parliamentary form of government through the actions of the people provided an example of the people "making good laws and building good institutions." Finally, our Founding Fathers had a range of beliefs which at the liberal end of the spectrum included Deists who were not strong church supports. Joseph Priestley, founder of the first Unitarian Church in America, commented on the beliefs of one Founding Father, Benjamin Franklin.

It is much to be lamented that a man of Dr. Franklin's general good character and great influence, should have been an unbeliever in Christianity, and also have done so much as he did to make others unbelievers.

The emergence of our democracy based on a secular Constitution provided a safe haven for those who were being religiously persecuted for heretical activities in other countries. Many came to America for religious freedom, and the country has benefited from the resulting religious diversity of the people.

One of the Founding Fathers, Thomas Jefferson, thought long and hard about religion and its intersection with governance and philosophy. He offered this advice to the new country about religion:

Fix Reason firmly in her seat, and call to her tribunal every fact, every opinion. Question with boldness even the existence of God; because if there be one He must approve the homage of Reason rather than that of blindfolded Fear. Do not be frightened from this inquiry by any fear of its consequences. If it ends in a belief that there is no God, you will find incitements to virtue in the comfort and pleasantness you feel in its exercise and in the love of others which it will procure for you.

The writings of Thomas Jefferson and James Madison were used as reference to the implementation of the separation of the state from religion as a foundational concept in the new American democracy. They first toiled to have their state, Virginia, free of religious involvement and then proceeded to effect the separation at the national level.

The prevailing civil philosophy of the delegates, notes Elizabeth Shakman Hurd, was Christian secularism: a political stance premised on a "chiefly Protestant notion of religion understood as private assent to a set of propositional beliefs." Molly Worthen explains:

> *The idea of Protestant civil religion sounds strange in a country that prides itself on secularism and religious tolerance. However, America's religious free market has never been entirely free. The founding fathers prized freedom of conscience, but they did not intend to purge society of Protestant influence (they had deep suspicions of Catholicism). Most believed that churches helped to restrain the excesses of mob democracy. Since then, theology has shaped American laws regarding marriage, public oaths and the bounds of free speech.*

Thus, although Protestantism was the base from which the laws separating church and state were written, these laws of the land are secular.

THE CONSTITUTION

Philosophers of the Enlightenment had argued that there could be a moral government without divine guidance as far back as Aristotle, who Patricia Churchland notes saw social morality as critical to governance:

> *. . . not as a divine business or a magical business, but as an essentially practical business. Making good laws and building good institutions, he thought of as cooperative tasks requiring intelligence and understanding and a grasp of the reverent facts.*

At one time during the deliberations of the Constitutional Convention, a preamble describing the new country as essentially a Christian country was discussed and placed before the delegates for consideration, but it failed to receive adequate support and was dropped, a secular preamble written and approved (the one we have now). Other discussions of Christian references were also rejected, and a concept of the separation of church from the State agreed upon and codified in the First Amendment to the Constitution:

Congress shall make no law respecting an establishment of religion, or prohibiting the free exercise thereof.

Our new secular democracy rejected God's authority over governing as outlined in the scriptures and replaced it with the authority of the People to govern, in essence replacing God's authority with the People's (Nature's) authority. The base authority of our Constitution is that:

there is no authority {to govern} except from the People {Nature}.

Our democracy gives citizens freedom of religion to whatever degree they wish outside of the government, but within our government, secular democratic laws are the laws of the land.

Our Founding Fathers were able to define the role of religion in the government. The result was that the Democratic Revolution in America was able to take practical and concrete steps to establish a secular democracy based on a Constitution that removed God and kings from government. Importantly, it was a majority of Christians who voted to establish the country as a secular one in which religion was separated from governance. Now, over two hundred years later, history has shown that our secular democracy has fulfilled much of its promise. In comparison to other governments, it has performed well, but not perfectly, in providing human rights and religious freedom for all of its citizens.

But the belief by some Christians that their God must be the governing authority and not the secular Constitution has remained since the founding days. At that time many Christian leaders were concerned with the non-inclusion of their God in the new government and some Christians attacked the Founding Fathers for their beliefs and for their views on the separation of God from the new government. A history textbook[47] from a Christian university quickly labeled Thomas Jefferson as a liar and Antichrist for his Deistic beliefs:

> *American believers must be aware of his (Jefferson) views of Christ as a good teacher, but not as the incarnate son of God. As the Apostle John said, "Who is a liar but he that denieth that Jesus is the Christ? He is antichrist, that denieth the Father and the Son."*

The powerful Timothy Dwight, president of Yale and a leading Christian, attacked James Madison, who was later to be elected president, for his support of the Constitution:

> *The nation has offended Providence. We formed our Constitution without any acknowledgements of God, without any recognition of His mercies to us, as a people, of His government, or even His existence. The {Constitutional} Convention, by which it was formed, never even asked once, His direction, or His blessings, upon their labors. Thus we commenced our national existence under the present system, without God.*

Other Christians predicted the failure of any state that did not include God and their Christian message. Religious organizations have rarely acknowledged (who knows of any?) that a secular democracy could provide a government with more freedoms for its religious citizens than a theocracy. The Vatican was not supportive to the American secular democracy over the first hundred years after its founding, for democratic precepts were at odds with the theocratic policies favored by the Vatican: (1) the separation of church and state, and (2) the forbidding of any religious test for public office. Even as late as 1864, Pope Pius IX declared in his *Syllabus of Errors* that:

> . . . *the Church ought {not} to be separated from the State, and the State from the Church.*

Theocratic governance has been a long tale of the repression of human rights by religious organizations. Theocratic culture and laws (God sends down the laws, and they are not contestable) have greatly limited the visibility into the governance (no one can question God's representatives over crimes), and the people have suffered. As with all organizations, including theocratic ones, having no voice by the People, no visibility by the People, and no power by the People leads to corrupt leadership and to the detriment of the people.

The Constitution is a secular document designed to guarantee personal freedoms, including the freedom of religion for all citizens. It is a living document that has been changed from time to time with twenty-seven amendments, including changes in the definition of a citizen.

When our Constitution was written, slavery was acceptable in the United States, as well as in the Bible and was the law of many countries. In 1864, during the Civil War, the heightened awareness of the evilness of slavery gave President Lincoln the opportunity to press for the Union to amend (Amendment XIII) the Constitution and abolish slavery. The Bible has not been corrected and still reads that slavery is acceptable to God.

The Constitution also harbored vestiges of other biblical and contemporary civil morals—the denial of human rights for women and homosexuals. Further changes in the Constitution have been made to reflect our evolving culture through Amendments: voting rights for citizens—for race and color with Amendment XV (1870) and for all sexes with Amendment XIX (1920). Other changes are still in flux, such as rights for homosexual citizens. The Constitution is a reflection of the slowly changing culture of our society. Recent polls[48] indicate that two-thirds of the citizens remain in favor of our democracy even with all of its faults.

Public education has been a major civil commitment in our country. With the use of taxpayers' money these schools adhere to the separation of state from religion. The science taught has been the secular science of the scientific community. However, Christians have tried and continue to try in a number of ways to prevent the secular Darwin's Theory from being taught in state public schools, including passing state laws prohibiting teaching it. Another approach by Christians has been to "water-down" Darwin's Theory by teaching religious theotheories as equals and arguing that it's just a theory and that supernatural theotheories should have equal consideration.

One blatant example of state intrusion into education from Christian pressure has been the Alabama state law that requires an insert to be placed into each high school biology textbook declaring Darwin's Theory as just a theory. This is being used by Christian biology teachers to teach theotheories, such as, biblical Creationism as a theory in place of or in parallel with Darwin's Theory. If an insert was written by scientists instead of Christian state officials, an insert on evolution would be far different. Appendix C: Key Tenets of Darwin's Theory: Summary Reference for Biology Students is included as a guide for those interested in a science replacement of the Alabama insert on evolution.

In general, our secular constitutional government has worked well, and public education by the government has largely kept religion out of the schools while providing freedom for all religions. Freedom from religious intrusion into government waxes and wanes with variations in the elective strength of Christians in the government. However, when Christians are in a majority, there is a constant pressure toward inserting the Christian God into government. Inroads into removing some of the wall separating religion from the government are made from time to time as Christians continue to push toward a "Christian Nation." These are the problems of maintaining a secular government in which no one religion is supported and all are treated fairly. Such problems were anticipated by our founding fathers. Thomas Jefferson noted that for our State:

The price of freedom is eternal vigilance.

CONSTITUTIONAL IRONY

Fundamental Christians still argue that the country would be more moral if it were a "Christian Nation." History presents a little irony with the "Christian Nation" argument. The Confederate States of American was a country formed from states that split from our country over issue of slavery. The Civil War that followed had the Confederates embrace slavery and appeal for guidance from the Christian God. In their constitution they included an appeal to God, "invoking the favor and guidance of Almighty God." The North retained its secular Constitution, which did not seek divine guidance, and fought against slavery. The country without the guidance of Almighty God won the war. Such is the value of godly guidance from a "Christian Nation."

CHRISTIAN PRESSURE

Early in our history, some Christians acted as if this was a Christian nation. A pioneer, freethinker, and Universalist, Abner Kneeland got into trouble in the 1830s by publishing[50] in a paper, *The Investigator*, one of his letters that said:

> *Universalists believe in a god who I do not; but believe that their god, with all his moral attributes, (aside from Nature itself,) is nothing more than a chimera of their own imagination.*

In Massachusetts, Kneeland was charged with blasphemy in 1834 for saying he did not believe in God, suffered three trials, was convicted in one of blasphemy, and served sixty days in jail. The prosecuting attorney for the Commonwealth of Massachusetts told the jury that if Kneeland was not punished "marriages [will be] dissolved, prostitution made easy and safe, moral and religious restraints removed, property invaded, the foundations of society broken up, and property made common." His appeal to the state Supreme Court concluded with a split verdict of guilty in 1838. It was the last case of blasphemy brought to a legal court in this country.

As to be expected in a democracy, the boundaries of laws will be continually tested, and in the case of the interface between religion and the state, this has certainly been the case.

Many Christian Americans assume that God has authority over science and that they should insert religious theotheories into public school science education (teach Creationism or ID) and into government research (limit on stem cell research). This is to be expected[52] in the general US population, where about a third (35 percent) believes in Darwin's Theory of evolution, while 48 percent continue to believe in supernatural biblical creationism. At the same time, belief by the US public in God and God's supernatural world has remained high—93 percent believe in a supernatural God, 84 percent believe in miracles, 66 percent in heaven, and 85 percent in angels. Christian religiosity in the population at large is one of the highest in the world for industrial countries and continues to be a barrier to the independence of science. The scientific community believes differently, with 80 percent saying they do not believe in a God, but they are a small percentage of the population.

The "Christian Nation" myth continues to surface with some Christians arguing that the United States was founded as one or that it should be converted to one. The maintenance of our democracy to be free from the authority of religion continues to take considerable effort.[49] Today, some Christians have forgotten (or most likely, refuse to accept) one of the founding fundamentals of our democracy—religious freedom. It is not clear that advocates for a "Christian Nation" know the difference between a democracy and a theocracy—they know not what they wish for. This is explored in *Jefferson's Scissors* by Perry.

In many ways Christians have resisted the democratic principle of the separation of religion from the government (examples include: organizing school-led prayer in classrooms, installing Christian icons on public land, etc.).

In some places around the country, Christians have erected monuments of the Ten Commandments and Christian crosses on public land, most of which have caused lawsuits to enforce the separation of church and state. One example, close to the author's home in California, has been the Christian cross erected on public land atop a hill called Soledad. Law suits and legal wrangling have continued for twenty-three years. The case made it to the Supreme Court who agreed not to hear a lower court ruling against retention of the cross on public land. The separation of church and state was upheld in this case, but the legal struggle continues for some Christians refuse to accept the concept of separation of church and state.

Another myth of the "Christian Nation" is that our laws are based on the Ten Commandments. This myth[51] has even been supported by a Supreme Court justice, Antonin Scalia, who obviously failed to notice that there are three versions of the Ten Commandments. He has not indicated which of the two to reject and why he would reject them. For example, the Ten Commandments in Exodus 34 states: "Dedicate the first offspring from every womb to God" and "Offer not the blood sacrifices with leaven bread." These do not sound like a sound basis for democratic laws in a diverse society today.

In a democracy, only four of the ten biblical commandments have any utility for civil law. For example, the first four of the commandments could not be used for laws in a secular country, for they are Christian God specific. The fifth (honor one's parents) and the Tenth Commandment (don't covet a neighbor's property) do not provide a sound base for enforceable secular civil laws. Commandments six through nine (prohibition of murder, adultery, stealing, and false testimony) are useful, as they have been used by many cultures, from Hammurabi's code to English Common law.

Those Christians arguing for a Christian nation forget (conveniently) that if one follows the biblical morals, they would have to convince blacks to be slaves again, women to give up the vote and many legal rights (equality with men), and homosexuals to accept legal persecution.

Our Constitution is the end product of a long struggle by peoples of many countries to throw off the yoke of theocracies. The Founding

Fathers foresaw this when they embraced freedom for all religions. Although the country did incorporate some biblical law (example, inferiority of women) in the writing of the Constitution, many of these have been changed or are in the process of being changed. Bishop Spong observes some of the consequences of biblical law in a democracy:

> If we literalize the Scriptures, as Christians have tended to do and which fundamentalists do without apology or hesitancy, we also literalize the prejudices of that era, which were against democracy, against people of color, against women and against homosexual persons.

One recent conflict[53] with a religious authority and medicine in our secular country is an example that touched all three: religion, science, and democracy. The background is a Catholic-administered hospital and a pregnant patient who had an emergency illness that threatened her life. The hospital's medical team assigned to save the woman's life said they would have to abort the fetus. The hospital's ethics council (which included a Catholic sister) met quickly and approved the procedure, for it gave the best chance to save the woman's life, and an abortion was performed.

The local Catholic bishop heard the story and excommunicated the Catholic sister on the medical ethics team for violating the Vatican's ban on abortion. The hospital refused to fire the sister or the doctor who performed the operation, saying they acted ethically and most humanely. Not getting its way on firing the sister or the doctor, the bishop then excommunicated the whole hospital, putting hundreds of patients, many of whom were Catholics, at risk. The hospital ignored the bishop and continued to provide its medical services to the community, but it no longer has Catholic services within the hospital.

In the end, the authority of the secular government's medical ethics guidelines was followed, and the authority of the church was rejected. This is a clear case of religion overstepping its authority and making medical decisions that science (what procedure to perform) and the state (who could get what treatment) should make. Sadly, this case undermines the major humanitarian services Catholics provide to thousands through the country. In some states, Catholics have the highest percentage of the hospital beds in the state.

THE IRISH EXAMPLE

Many examples have pointed to the fact that mixing God with the state (government) and Nature (science) produces an unhealthy contamination of all three. The two thousand year history of Christian theocracies is full of stories of lost freedoms and roadblocks to modernity. Fortunately, in America the Democratic Revolution did occur, and the separation of religion from the state was written into our founding Constitution.

Other countries with long histories of theocratic governance, however, have continuing problems over removing the church's authority in civil governance. Ireland is an example of a state striving to establish the authority of their state over the unwholesome reach of the church into the civil protection of children's well-being. The public unfolding of tragic Irish clergy pedophile cases has highlighted the reluctance of the Catholic Church to give up on its long held theocratic position of elevating canon law above civil law and ignoring state laws. The church has resisted supporting the secular state laws for all of the people and religious institutions, although public safety has been seen to improve when laws based on biblical morals used by the church have been replaced with secular state laws.

Over many centuries the Catholic Church was essentially part of an Irish theocracy. Ireland was the most religious of all European countries—effectively, a Catholic theocracy. Sarah Lyall[45] summed up the situation:

> *This is still a country where abortion is against the law, where divorce became legal only in 1995, where the church runs more than 90 percent of the primary schools and where 87 percent of the population identifies itself as Catholic.*

The need for the government to be independent of religion finally came to be a glaring issue in Ireland after the church gave only lip service to protecting the civil rights of children. A *New York Times*[46] report on the statements by the Irish Prime Minister Kenny tells the story:

> *After 17 years of revolting revelations, Kenny said the latest report on the Cloyne diocese in County Cork exposed "an attempt by the Holy See to frustrate an inquiry in a sovereign, democratic republic as little as three years ago, not three decades ago."*

> *The report, he said, "excavates the dysfunction, disconnection, elitism, the narcissism that dominates the culture of the Vatican to this day." The rape and torture of children were downplayed or "managed" to uphold, instead, the primacy of the institution, its power, standing and "reputation."*

Far from listening to evidence of humiliation and betrayal with St. Benedict's "ear of the heart", the Vatican's reaction was to parse and analyze it with the gimlet eye of a canon lawyer. This calculated, withering position was the polar opposite of the humility and compassion upon which the Roman church was founded.

Pulling back the curtain to expose the profane amid the sacred would have been remarkable coming from any leader in one of the many countries scarred by pedophile priests, but from the devoutly

Catholic prime minister of a nation whose constitution once enshrined the special position of the church, it was breathtaking.

What else needs to be said to argue that theocracies or governments that excessively mix religion and state do not provide good government for the people? This example emphasizes the need for the separation of church and state. The checks and balances provided by an independent government and science are essential. America included the separation of church and state in its Constitution in 1778; Ireland is now beginning to follow the democratic example of separating the church from civil powers. The extensive powers exercised by the past Catholic theocracy die slowly and with great pain to all, as the transition of Ireland from theocratic power demonstrates.

God's Way

The beautiful vistas and the sweeping diversity of life observed by Naturalists are also viewed by Christians and those of other religions. Where Naturalists see a world of Nature's workings, Christians see these as God's gifts to man. Further, Christians feel that they are supported by a community of believers in their views. Gary Gutting's observation on religious knowledge gives us a start for defining God's way:

> A religion offers a community in which we are loved by others and in turn learn to love them. Often this love is understood, at least partly, in terms of a moral code that guides all aspects of a believer's life. Religious understanding offers a way of making sense of the world as a whole and our lives in particular. Among other things, it typically helps believers make sense of the group's moral code. Religious knowledge offers a metaphysical and/or historical account of supernatural realities that, if true, shows the operation of a benevolent power in the universe. The account is thought to provide a causal explanation of how the religion came to exist and, at the same time, a foundation for its morality and system of understanding.

In the supernatural Christian narrative, God is described as a miracle maker, an omnipotent force, and an Ex-Nature God from out of this universe who created the universe and man. To be a Christian is to accept, literally or metaphorically, biblical supernatural knowledge and believe in God's special relationship with and authority over man and Nature. From a Christian belief viewpoint, it follows that the answer to the question posed, *Who to thank for evolution?* should be taken from Genesis—that is, *thank God for creation.*

But the question is about evolution, not creation. Evolution is a biological process of Nature that became a science theory to be reckoned with by Christians after Darwin introduced it 150 years ago. Before that and for two thousand years, *thank God for creation* had worked quite well for Christians. But now some Christians want their God to also *thank evolution* as well as *creation* described in Genesis. This causes a conflict with the view from Nature's world on the creation and design of the universe, the Earth and man. This section presents the growth of Christianity and it's conflicts with the natural science processes of the creation and evolution of the universe and man. Arguments by Christian authors supporting the view to *thank God for evolution* are presented.

EARLY RELIGION

As man evolved, beliefs arose to explain the mysteries of why he was punished at times and at other times helped by mysterious forces out there, as well as by members of his tribe. Man is naturally inclined to seek the cause of events. Robert Wright notes the search for answers:

> *When bad things happen to you, it often means someone is mad at you, maybe because you've done something to offend them; making amends is often a good way to make the bad things stop happen-*

*ing. If you substitute "some god or spirit" for "someone" you have
a law that is found in every hunter-gatherer religion.*

Within early tribes spiritual activities addressing the mysteries
were guided by shamans, individuals who put themselves forward as
leaders to explain the mysterious spirits, omens, and gods. In primitive religions there developed a deep reverence for shamans and superstitions, as Wright notes:

> *Because whenever people sense the presence of a puzzling and
> momentous force, they want to believe there is a way to comprehend
> it. If you can convince them that you're the key to comprehension,
> you can reach great stature.*

Shamans of early primitive hunter-gatherer societies offered explanations for events, counsel on actions to be taken, and direction to receiving wisdom on the future. By providing such service, shamans were elevated by their peers to positions as powerful leaders in their tribes. It is argued that cohesion through common beliefs enhanced the survival of the tribe. Playing on the fears of the unknown, shamans were able to expand their personal power, and their advice was broad, ranging over many personal as well as tribal governing decisions. Wright observes:

> *The shaman represents a crucial step in the emergence of organized religion. He (or she, sometimes) is the link between earliest religions—a fluid amalgam of beliefs and spirits—and being a shaman was the first step of leadership to those today we call archbishops or ayatollahs.*

Shamans serviced this need by supplying rituals, group morals, and mysterious gods that took many forms: animals and objects in their environment; sun gods, moon gods, etc. Over time they became parts of emerging religions. Ancient archaeological artifacts from burial sites indicate man's attempts to think about the supernatural (gods and an afterlife?), spiritual, and religious activities. Religious icons started to appear (about 50,000 BCE), indicating that religious rituals possibly added to the tribe cohesion, which aided survivability and growth. Decorated burial sites, a thirty-seven-thousand-year-old flute, and cave paintings indicated early expressions of the arts, music, and spirituality (to an unknown degree).

Religions have emerged from many cultures, and although they may differ in detail, they have a common element of embracing the supernatural. Although there are other common elements, religions can be quite varied. Jared Diamond gives an overview with fifteen definitions of religion in his book *The World Until Yesterday*. An example of one such definition by Daniel Dennett is:

> *Social systems whose participants avow belief in a supernatural agent or agents whose approval is to be sought.*

Sites for religious activities have been found that date back to pre-urban days, such as a site[11] being excavated in Turkey dating back over twelve thousand years that indicates organized religion may have predated farming and villages and may have played a significant role in urbanization. But there were many things happening, for this was also the time that dogs and wheat were domesticated and farming and urban life began.

In the area of Jericho, settlements date back to about 8,000 BCE. About 5,000 BCE, man invented metal tools and weapons and started to wall his community for defense. Larger communities, like city-states, began to appear. Technology continued to advance after settlements became a way of life with the invention of the plow and improvements in copper smelting that led to better metal tools.

Over the next three thousand years, urbanization expanded, and other metals such as tin, bronze, and iron appeared, and from this new technology new tools and weapons were forged. As societies became more stable and larger through farming and urban building, religious beliefs became more structured, and in time, Robert Wright notes that:

> . . . *religion came to be a distinct body of belief and practice, kept in shape by an authoritative institution.*

The invention of writing (about 4,500 BCE) first appeared as cuneiform writing on clay tables in Babylonia and increased the capability for extended commerce by increasing the efficiency of communications and record keeping. The written codification of acceptable (and unacceptable) behavior in the form of social laws increased the stability of the community with laws (moral and religious rules). Hammurabi's laws, dating from 1,750 BCE, are an early example of societal-religious laws for community governance. Such laws were an important step in the growth of the community, from tribes to chiefdoms to mighty kingdoms.[12]

JEWISH HERITAGE

More complex and organized tribes with new religions appeared, including Judaism about 1,500 BCE. The historian John Bright[13] notes that the Hebrew were latecomers on history's stage in the Middle East. All across the biblical lands, tribal cultures had blossomed and run their course for thousands of years before Abraham appeared.

In the Palestinian region, over fifty thousand years[54] ago, several pagan tribes occupied the land. About 2,000-1,500 BCE, Abraham[15] and his tribe came to this land after journeying from northern

Mesopotamia, and there, under the patriarchs Abraham, Isaac, Jacob, and Moses, the Hebrew tribes grew and began the transition from polytheism to monotheism. It took hundreds of years for the transition to one god.

Under King David, about 1,000 BCE, the tribes consolidated into the kingdom of Israel. The writing of the Jewish Bible from oral recitations began about this time and continued through the Babylonian captivity (586-538 BCE) and the return to Israel after Babylonia fell to Persian forces. The Hebrew narrative was written by several authors over this period, and the canonization of the texts occurred in 90 BCE. Biblical archaeology has been successful in coupling some of the early Israelites' towns and wars with the surrounding Egyptians and Babylonians. Some of the earliest evidence has been found of an Egyptian war with the occupants of Israel in 1208 BCE and the existence of the kingdom of David about 1,000 BCE.

The book of Genesis expresses the central theme of God creating man and appointing him regent over the Earth. Biblical scholars believe that Genesis was written to give an identity to the Israelites. Many biblical scholars have concluded that it was not written as a book to explain how God created man or the universe, but to declare their single God was the mighty creator.

CHRISTIAN HERITAGE

When Jesus appeared most Jews concluded that Jesus was not their Messiah, and they failed to join the early group of believers. To others Jesus was the Messiah and an exemplary teacher. The biblical narrative in the Old Testament describes the growth of Judaism, and later the New Testament adds the Christian growth until the death of the Christian Messiah, Jesus.[14]

Over the first few hundred years after his death, his followers reported his words and life and expanded the Christian belief so
that it became a separate religion from that of its Jewish roots. Jesus
extended man's morals and humanitarian outreach beyond the Jewish
tradition. His moral teachings and hope for salvation provided the
foundation for the new religion, Christianity.

With the appearance of Jesus, the humanist-savior Messiah,
Christianity began its journey to become a new religion separate
from its Jewish roots. The removal of some Jewish religious requirements, such as circumcision and dietary restrictions, appealed to new
prospects and helped fuel the growth of the new religion. Christians
focused on servicing the human condition, humanity, and salvation,
and offered a different religious perspective with the view of a mysterious messiah god who offered guidance during life and salvation
after life.

The stage set by the early Christians to define their religion. Early
on, the Gospels of Mathew, Mark, Luke, and John formed a base for
the new religion, but much remained to be done to get it on paper. A
view on the sequencing has been expressed by Bishop Spong:[55]

> *The gospels were not written by eyewitnesses. They are the products
> of a time between two and three generations after the crucifixion
> of Jesus. The gospels were written in Greek, a language neither
> Jesus nor his disciples could either speak or write. We can find no
> evidence that miracles were associated with the memory of Jesus
> prior to the 8th decade. The stories of Jesus' miraculous birth to a
> virgin did not enter the developing Christian tradition until the
> 9th decade.*

Two generations after the death of Jesus, his followers began
documenting the new religion in the New Testament, essentially an

anthology, a collection of works written at different times by various authors dealing with first-century Christianity. It would take three hundred years for Christians to agree on a base orthodoxy.

The center of the New Testament, the Gospels, consists of four narratives of the life, teachings, and death of Jesus. The original texts were written beginning around AD 40, and the last writings were incorporated in the mid-second century AD. The Christian Bible also included the Hebrew Bible with its miraculous supernatural events, such as the creation of the universe and man, creating man in God's image, a stationary star over Bethlehem guiding wise men, Noah's worldwide flood, lightning bolts sent by God, and a whale swallowing Jonah.

In the past, the writing of Genesis was attributed to Moses, but now biblical research points to others. Although several writers have been identified, there is still an ongoing scholarly debate[56] on who they were. Richard Elliott Friedman has summarized the research on the authors of Genesis in his book *Who Wrote the Bible?* He notes:

> *The Bible is thus a synthesis of history and literature, sometimes in harmony and sometimes in tension, but utterly inseparable.*

Many Christians have looked for proof of the biblical miracles, but none have been forthcoming. Some biblical scholars say that since these events were intended only as metaphors expressing the might of their God. Literal minded believers continue to search.

One event that has received many investigations is the global flood account, but no direct evidence has been found and what has been learned argues against it. What scientists have found instead is a local flood caused by the Mediterranean Sea flooding inland and creating the Black Sea about seven thousand years ago. Such a flood would have been a disaster to the villages bordering on the Black

Sea, but not to the rest of the world. The occurrence of this flood was at the beginnings of major civilizations in Mesopotamia. Here the flood story became folklore of the region, and memories of it were kept alive verbally in epic tales. When writing was invented in about 4,000 BCE, the flood event was recorded. But by then it had become a worldwide flood tale. Epics written at the time, including the epic of Gilgamesh and the Sumerian gods, all included a flood tale. From there it is easy to understand how several thousand years later the same flood tale was incorporated by the Jews into their Genesis narrative. Thus, a natural event, a local flood, had over thousands of years become a mythical global flood.

The Christian mythos, or the spiritual side, has had great appeal to many. Jesus taught an appealing moral perspective: love thy neighbor and help the poor. Christians believed that he also sacrificed his life for man's sins and promised eternal life for those pursuing a moral Christian life. This has proven to be an inspiring God to seek, to worship, and to follow and the appeal to the poor and downtrodden continues to be successful. Organizations within the church were established early to address the poor and the disenfranchised, and many of these organizations continue to this day, providing needy services to humanity. The Christian mythos has been and remains a vital force for humanity's good.

CHRISTIANITY'S GROWTH

Christianity started out as one of several religions in a territory dominated by the mighty Roman Empire. Within a few hundred years of its birth, the Roman Emperor Constantine gave it a major boost when he added, for the need of more soldiers, Christians into his armies. He later legalized Christian worship within the empire alongside Pagan worship. Within a relatively short time, Christians gained statue in the government, learned to be a political force, expanded their power, and became a major player within the Roman theocracies.

Early in 325, Emperor Constantine made an effort through the Council of Nicaea to bring a unifying identify to the Christian factions and settle the controversy over the relationship of Jesus to God the Father. This issue was difficult to settle and remained unsettled until other councils, primarily the Second Council of Nicaea in 787, known as the "Triumph of Orthodoxy."

A dark side of the Christian growth was their use of civil power within theocracies to brutally eliminate Pagans across the Roman Empire in order to solidify their power base. It was during this period (around 400 to 500 CE) that the church began to impose its control over peoples, including scientists and philosophers, within the empire. The church simply went on a campaign to eliminate the opposition that did not join or convert. This earthly power play was a clear example of natural selection of religions at work and a poor example of a godly action by a benevolent God administrating to his flock.

One example of the Christian campaign of violence in their early growth period was the actions directed toward Pagans, specifically Hypatia, a woman, a Pagan, and a leading scientist and philosopher living and teaching in Alexandria. She taught openly to students of all religions and was a highly visible Pagan, making her a thorn in the side of the emerging Christian leadership. In 415 she was savagely killed by a Christian mob in the streets because of her fame as a teacher and a Pagan among the youth. Her death and fighting among Christians and Pagans led to the burning of the serapeum, a Pagan temple, in the great library of Alexandria. Other Pagan centers of learning were destroyed in many other cities in the Roman Empire. It was a sign of Christian power that would include scientists and philosophers, for all Pagan subjects of the Roman Empire were to convert or be killed.

THEOCRACIES

Within five hundred years after its formation, the Christian church became the dominant religion in the Western world, and for a thousand years afterwards it retained that lofty position. Christianity, through its power in theocratic governments, held that God was the authority not only over religious matters, but also over state governance and science. From its position of vast power centered in the Vatican in Rome, it established dogmas in religion, science, and governance, by which it governed its theocratic world.

But all large and powerful organizations managed by fallible humans, including the religious theocracies and churches, are fertile fields for human corruption and the ills of large organizations. The Catholic Church was not an exception, and after years of governing, corruption grew and in many areas became widespread.

As the church grew larger, corruption became broader and deeper through the sale of indulgencies and other forms of corruption. Throughout the empire, various theologians and members of the Church argued for changes in the dogma and an end to the sale of indulgences. One early forceful voice was that of Jan Hus, who voiced his opposition to the dogma of the sacraments and to indulgencies, a tax on belief. For these he was burned at the stake in 1415 for the church chose to defend its dogma and not reform. His followers were outraged, and after rioting started, the pope sent a crusade of Christian soldiers to squelch this uprising. The people continued their outrage, and the confrontation turned into a war that lasted over ten years.

A century later a second powerful voice, Martin Luther, was heard in 1517 when he nailed a paper (Ninety-Five Theses) with his objections to indulgences to the church door in Wittenberg. His refusal to retract his writings at the demand of Pope Leo X in 1520 and the Holy Roman Emperor Charles V at the Diet of Worms in 1521 resulted in his excommunication by the pope and his condemnation as an outlaw by the emperor. Luther survived these attacks by protection from local princes in several states. Luther's actions led to the widespread Protestant Reformation, which split the church between Catholics and the emerging Protestants.

It was in this unsettled environment that the church faced sustained challenges to their authority from three sides: in religion with the Protestant Reformation challenging the very meaning of being a Christian and the authority of the church, in science from the new secular theories challenging entrenched dogma, and in governance with the secular Enlightenment challenging the philosophy of the church in morals and laws.

A leading Enlightenment philosopher and writer was Denis Diderot, who preached the right of the individual to determine the course of his or her life. Andrew Curran summarizes:

> *His message was of intellectual emancipation from received authorities—be they religious, political or societal—and always in the interest of the common good. More so than the deists Voltaire and Rousseau, Diderot embodied the most progressive wing of Enlightenment thought, a position that stemmed from his belief that skepticism in all matters was "the first step toward truth."*

Modernity was knocking on the Vatican's doors. Christianity had a marvelous growth from its inception to the lordship role in almost every country in the West. It was at this pinnacle of growth that the advances in philosophy, science, and governance confronted the dogmatic commands of the church. Diderot and others wrote articles banned by the church, and in others, Curran notes that:

. . . he challenged arguments supporting the existence of God in "Letter on the Blind"—a book that earned him three months in prison.

In explaining what makes churches tick; E. O. Wilson has noted that religions and their institutions (churches) are products themselves of evolution. Religions are:

'. . . a kind of cultural Darwinism operates during the competition among the sects in the evolution of the more advance religions. Those that gain adherents grow; those that cannot disappear. Consequently, religions are like other human institutions in that they evolve in directions that enhance the welfare of the practitioners.

Wilson observes an emotive and social power in religions, and for these reasons, one can see that churches are social organizations that supply membership in a group by offering support for physical, psychological, and spiritual needs. Christians have simply outperformed and gained more power than their religious competition in many areas of the globe.

CHRISTIAN TRUTHS

In the thirteenth century, Thomas Aquinas presented arguments for two Christian truths, revelations and reason. From this, Descartes in the seventeenth century further defined Christian theology through the concept of dualism—the truth about the mind is supernatural, and the truth about the brain/body can be determined by its natural material properties. Dualism assumes that natural phenomena can be discovered, but man's design was considered to be different, supernatural, and not subject to questioning. The dualistic view was seen by theologians as the product of an underlying supernatural plan by God. Theologians believed that with God's authority over religion, science, and the State, all evolved under God and could exist peacefully.

The dualistic view of truths became central to the Christian view of Nature. Dualism does not prevent the study of Nature, but allows supernatural explanations that are beyond Nature. This position was explained by William Paley in 1802 in his *Natural Theology*. Nature was seen by Dualists as the product of God's supernatural plan, which can be neither directly studied nor understood. Greg Graffin notes:

> *The dualist easily segregated knowledge into material and supernatural without challenges of unifying the two spheres. Problems arose when scientific pursuits revealed verifiable truths that conflicted entirely with traditional theological explanations. This difficulty was cleverly minimized because a dualistic perspective always allows for new discoveries, new additions to natural knowledge, while still remaining a belief in entities beyond the scope of traditional scientific methods (the supernatural "truth by revelation").*

Darwin's Theory directly challenged the dualistic view, and with the subsequent acceptance of Darwinian evolution, dualism is no longer seen as a viable argument. Dualism can no longer provide a cover for theology to peacefully coexist with science. Darwin, by declaring that God's supernatural world was unnecessary for explanations of living organisms, established that Nature's world (Monism) was an adequate base[57] for understanding. Greg Graffin notes:

> *Monism spread within evolutionary biology progressively with Darwin. It is the logical philosophical antithesis of dualism. Where a Dualist sees permanent mystery and limitations of knowledge in the sphere of the supernatural, a Monist sees a vast*

universe of exploration in the natural world, parts of which we have not had the time yet to comprehend. To a Monist, all natural phenomena are the result of evolution, which entails nothing but Nature's law to explain their origin in material terms. The supernatural has no explanatory purpose.

Monism sets the stage for the independence of science from religion. However, the transition is not complete, for some Christians continue to support the dualistic view of God's supernatural claim over the authority of Nature and the State, even though this view has been upended by advances in natural science.

Christians have mounted defenses against the removal of their God's authority over natural science by proposing theotheories, replete with miracles, to justify their beliefs in biblical literacy in the face of the advance of science. Two examples: (medicine) the church declared epilepsy to be caused by demon possession and (science) the earth was declared to be at the center of the universe. As science progressed in the natural world, epilepsy was proven to be a medical condition from which the sufferer experiences physical seizures from natural causes in the brain. Placing the sun at the center of the solar system with the earth orbiting the sun matched observations from astrometry. Epilepsy's natural causes and the helio-centric solar system's natural origin have been subjected to verification by testing. Supernatural demon possession in man and the earth-centric theotheory in the heavens have been rejected.

LORDSHIP PROBLEMS

By the Middle Ages, Christian theocracies had become the authority and power for setting rules and laws not only for religion, but for natural science and government in the Western world. The Vatican was run as a monarchy in which the pope was king reigning over fiefdoms.

But the rising intellectual revolutions of the Enlightenment and the scientific revolutions brought challenges to the very basis of power of Christian theocracies. It was the philosopher Bernard Spinoza in 1640 who first questioned the authority of religion to rule with secular arguments. The Democratic Revolution that followed in a hundred and fifty years in America was built on the Enlightenment's arguments, which questioned the validity of the extent of God's authority over governance.

As the Christian Church grew in power and influence, the institutional side of Christianity started demonstrating many of the less appealing human aspects of organized religion—the self-granted right to control many aspects of believers' lives beyond their religious needs and the right to convert non-believers to believers, many times at a terrible cost of human life. The church's view of justice for those who challenged the church dogma was to declare them heretics, for that gave them the right to torture, force guilty confessions, and conduct trials which were used as spectacles to frighten others. They could do this, for the Vatican supplied the laws, the court, the prosecutors, the jailers, the jury, and the judge. Confessions from torture and the final outcomes for heretics were never in doubt. The church was in charge of everything—so they thought.

As dissidents, or heretics, challenged the dogma grew in numbers, the Catholic Church established a policing organization, the Inquisition, in 1184 to keep the purity of the faith and the power of the church dogma from the increasing pressure of growing numbers of discontented believers within the church, from small Christian spinoff groups outside and from scientists, both within and without the church. These dissidents, skeptics, and reformers of the church's dogma were widely investigated, accused, and persecuted. The Vatican theocracy ruled by a combination of religious and civil authority and continued to exercise near total authority over scientists, people, and governments for over a thousand years.

In many cases, the church expanded into areas (science and governance) where Jesus (in opinion of some Christians) would not have entered, for such areas offered little positive spiritual ground for believers, only grounds for material power for the church bureaucracy. But this lofty position of lordship by the church bureaucracy was to be shaken by advances made by scientists, philosophers, and theologians, which led to wide-spread revolutions in intellectual thought.

Up until about 1500, the Vatican's assumed authority over science and governance was rarely challenged, and when challenged, it had the theocratic power to enforce its dogma by force, including death or imprisonment. But all this changed with the revolutions outgrew the ability of the church to suppress the advances: in science, starting with the Copernican in 1543, from the Enlightenment starting in the late 1700s, and the Democratic Revolution in 1789. These was followed by Darwinian Revolution in 1859 and the Neuronian in the 1990s. These revolutions have seriously challenged the Christian authority, or lordship, over natural science and state governance.

Struggles by the Catholic Church to keep up with the advances in natural science and how to control them have varied with the different views of the popes. By 1925 Pope Pius X lamented that sacred studies were being neglected because the study of natural sciences consumed too much time.[58] A later pope in *Humani Generis* denounced those who interpret biological evolution as a total theory of human origins and declared that only God could create the human soul.

As the pace of scientific studies increased, this religious position of Christian churches became more difficult to impose on science classes in the public schools. Top-down control of science by church fiat was no longer possible as the scientific communities became larger, more widespread, and interconnected. Finally, in the US public schools separated the science from religion and religious control was lost.

KNOWLEDGE REVOLUTIONS

Revolutions both in science and governance have greatly impacted the understanding of the relationships between God, Nature, and the State. Christians continue to struggle among themselves to understand the relationship of their biblical narrative with science and governance. Biblical inconsistencies about events, such as the order of creation in Genesis, where God make the grass and trees a day before making the Sun, were apparent to the early theologians of the church. For example, Saint Augustine (354-430 CE) noted:

> *It is also frequently asked what our belief must be about the form and shape of heaven, accordingly to Sacred Scriptures... Such subjects are of no profit for those who seek beatitude. And what is worst, they take up very precious time that ought to be given to what is spiritually beneficial to salvation.*

To Saint Augustine, the science (form and shape) is not what the scriptures are all about; they are about leading believers to individual spiritual salvation. He was trying to tell believers not to emphasize the shape of heaven, but to focus on the morals, ethics, and salvation offered by their religion.

Even when religion falls short of competing with scientific knowledge, religion can be an important alternate source of understanding humanism offered by their supernatural narrative. But there are many religions with supernatural narratives, and all do not agree. That is important only when one wants to find an absolute moral across all religions, but it is not a problem when one stays with one religion.

For a thousand years, Christian dogma was the controlling force before major revolutions in human thought, driven by modernity, arose and presented major challenges to the long-held church dogmas. As previously described, the authority of the church was reduced with the Copernican, Darwinian, and now the Neuronian revolutions in science. In governance it was the Democratic revolution in America that separated religion from government.

The understanding of Nature's worldview and its insistence on independence has created many challenges to religious dogma. Religious ideological and bureaucratic resistance to modernity continues and is ever present. However, some Christian leaders have moved to a broader acceptance of science and consider its role in the modern world. Pope John Paul II noted:

> *Science can purify religion from error and superstition; religion can purify science from idolatry and false absolutes. Each can draw the other into a wider world, a world in which both can flourish.*

This is an insightful view of science that can "flourish" along with religion. Further, John Paul II addressed the relationship of science in light of religion and stated that the Bible is not a textbook, so its discussions about science are made:

> *. . . not in order to provide us with a scientific treatise, but in order to state the correct relationship of man with God and the universe.*

This statement reaffirms those made by St Augustine many hundreds of years before. But there are resistors among Christian church leaders, for there are those who continue to press God-centric views of natural science theories on creation and evolution in today's world.

While John Paul II addressed his comment that science was an important discipline to consider in this modern world, other popes reverted to emphasizing the church's authority over science. This was illustrated in a 2010 statement by Benedict XVI in which he specifically notes that science is but a tool of God and reiterates the authority of religion over science:

> *The role of science is to reveal God in the universe: "Scientists do not create the world; they learn about it and attempt to imitate it," he said. "The scientist's experience as a human being is therefore that of perceiving a constant, a law that he has not created but that he has instead observed," the pope said. That perception, in turn, "leads us to admit the existence of an all-powerful Reason, which is other than that of man, and which sustains the world."*

It is obvious that this pope was attempting to assert the authority of the church (all-powerful Reason) over science, for surely he knows that secular scientists do not set out to devise experiments to reveal the Christian God or the Hindu god or any other god. That's a fine statement in a church, for it is a restatement of the omnipotence of the Christian God by a Christian, in this case a pope. However, the pope's statement was addressed to the wider world outside of his religious realm. The conflicts he raised are: first, is it the role of science to reveal God or to explore Nature? Second, does the pope have authority outside of his religious discipline to tell science what its role in the natural world should be?

But in other areas, such as the humanities, religion does have a role to contribute to our wholeness. Man's morals are from our evolved moral sense and experiences and can benefit from our religious beliefs. Ayala notes:

We humans have systems of morality concerning the consequences of our actions in regard to others, and derive meaning and purpose from religious beliefs.

Religion's contribution to morality, according to Ayala, is:

Religion deals with the meaning of life and the purpose of life, and the moral values that should govern our lives.

The intellectual revolutions over the past five hundred years have established Nature (science) and the State as independent disciplines alongside religion, as independent disciplines. Individuals, including Christians, have made major contributions to the advancement of science and have helped to define the boundaries between the three disciplines. Yet many Christian churches have had difficulty adjusting to the fact that their religion's voice is but one of a hundred religions, and religion is but one of three independent disciplines serving man: science (Nature) addressing natural world, religion (God) serving the humanitarian and religious needs of believers, and our democracy (the State) supplying a secular vehicle for citizens to prosper and grow and enjoy the benefits from science and religion.

Care should be taken to ensure that each of the three voices are heard. Indeed care should be taken to include the best from not only religion, but also science and our democracy.

SUMMARY

The strengths of Christianity—humanitarian services, morals, spiritual guidance, and salvation—have attracted believers to follow God's way, fueled its growth, and established its importance to the lives of many for two thousand years. This remains the case today. These brief notes on Christianity do not cover the breadth of, nor do justice to, the value of religion for believers, for the discussions are focused only on understanding the need for the separation and independence of religious supernatural beliefs from natural science and governance.

As outlined in the section Nature's Way, science advanced before, during, and after Christianity came along. Natural science has not been, nor is it now—and we will discuss this later—one of its strengths of Christian belief, although many individual Christians have made significant contributions to science. Since Darwin, some Liberal Christians have wished to appear scientifically knowledgeable and to believe that the insertion of natural science processes into the supernatural Christian narrative is necessary. This example—mixing evolution with religion has been tried, and the results from attempts to modify Darwin's Theory and combine it with the Christian narrative have been unsuccessful.

The Christian God's authority over his supernatural narrative is not questioned by science, nor by the State, for a person's belief in a God is a personal freedom. Science does not seek to prove that there is or is not a god; it just does not address gods, for the supernatural is to be avoided. What is questioned, however, are efforts by Christians to extend their God's authority over natural science, including evolution, and governance in the natural world. When this happens, science, as should be expected, comes out defending its independence, and indeed critiquing leading Christian authors has identified cases of unacceptable intrusion of religion into natural science.

Thank God for Evolution-Dowd

Michael Dowd's book *Thank God for Evolution: How the Marriage of Science and Religion will Transform Your Life and Your World* is a sermon (in the best sense of the word) by a minister selling God's "Good Word" as he sees it, which beckons all to believe that his proposed "marriage" of science and religion will remove all conflicts between them, transform your life, and bring harmony to the world. Many people who want such a "marriage" to work at any cost have accepted and endorsed his sermon.

Unfortunately, there are problems with his rosy "marriage" proposal: it just simply does not work in either the natural world of science or in God's supernatural world, for as a proposed unifying concept, it is scientifically and theologically unsound. Dowd rejects the metaphorical views on Genesis from biblical scholars and proposes an interpretation, a theotheory (I'll call Sacred Innovative Design), that is partly religious and partly scientific. Pursuing a "marriage" with this invented theotheory of evolution only leads to more conflicts, not the harmony and peace promised.

Dowd's book title exposes the problems he faces when attempting a "marriage" between religion and science. To *Thank God for Evolution*, he rejects the two leading views of the Genesis story: a literal description of God's story of supernatural creation supported by Fundamentalists and a metaphorical creation myth exalting God's power by progressive Christians. Instead, his Sacred Innovative Design theotheory of evolution is a modification of Darwin's Theory, with God directing evolution as the intelligent designer. Dowd then thanks his supernatural God for something that God never discussed in his Holy scriptures—evolution. The scriptures only cover man's creation, but Dowd proceeds to tell God to revise Genesis—drop your concept of man's creation and use Sacred Innovative Design instead of Darwin's Theory. Surely God must have asked Dowd, "Why are you asking me to change course and add something I have never considered nor needed to communicate with my believers? What is wrong with my supernatural Genesis narrative, which has successfully served the faithful for two thousand years? I am discussing salvation in Genesis, not science, so why should I allow the scriptures to be changed by the latest science theory?"

Dowd is not only making God unhappy, but scientists as well. In the natural world, scientists say you cannot insert a supernatural god into a natural theory, for any supernatural insertion would make the theory invalid. Further, the godless theory has worked quite well since Darwin proposed it over 150 years ago. Dowd tosses aside key tenets of the Christian religion and science as he attempts to marry the two in order to have God hitch a ride on Darwin's Theory under the presumption that God has authority over Nature. One can only conclude that Dowd's *Thank God for Evolution* does not make sense from either a theological or a scientific point of view. His subtitle, *How the Marriage of Science and Religion Will Transform Your Life and Our World,* outlines his arguments for his invented Sacred Innovative Design theotheory to replace Darwin's Theory and the biblical Genesis story. Dowd declares:

> *Once we accept that God's Word is not confined to ancient texts, and that God is still speaking through the public revelations of science . . .*

With this sweeping and unsupported assertion of his belief, essentially a rewrite of Genesis, Dowd is asserting that his supernatural God speaks through godless natural science. Dowd does not explain why God would want to speak through natural science, which is uncertain of its direction and changes its theories from time to time. Why did God not choose Darwin, Einstein, Crick, or others to speak his message on science in the past?

In constructing his Sacred Innovative Design theotheory, Dowd's first step is to give God the authority over Nature. To accomplish this Dowd begins by outlining two *I believes*, each with an accompanied *I know*:

> *I believe that God is creator and ruler of the universe. And I know that this statement is metaphorical, not literal, in what it says about the nature of reality.*

> *I know that God has been communicating faithfully, and clearly, for hundreds of years to the entire human community through the full range of sciences. And I believe that this has everything to do with fulfillment of the Gospel and Realizing Christ's return.*

These sweeping belief statements imposed on natural science may work for many Christians, but Dowd takes it even further by saying that godless science is a public revelation "to the entire human community." When Dowd says that his Christian God works through science, it is a rejection of what biblical scholars are saying and a religious assertion rejected by most Naturalists. Scientists would argue that science is not a transmission pipeline for the Christian God's views, or the views of any of the other thousand gods available. Most scientists would argue that Nature's science theories works quite well without the involvement of any God-centric theotheory. Most scientists avoid looking into the supernatural world of God, for they have enough trouble making things work in the natural world. Not many scientists would agree that they can see God's direction, nor would an Atheist, Hindu or Buddhist scientist agree, as Dowd claims, that science gives:

> *the fulfillment of the Gospel and Realizing Christ's return.*

Dowd's second *I know* is equally confusing as to why God would start communicating to humans only a few hundreds of years ago. Does this mean that God did not think much of Copernicus, Galileo, or Newton? If Dowd maintains that God was speaking through the science proposed by Galileo and not through his Christian church, the Vatican, why was Galileo found guilty? If God was speaking through Newton then why did Einstein have to come along and correct his work?

In the last two hundred years, God's communications have not been clear to many scientists, for in those years they have increasingly pointed out that their scientific work does not involve God. J. B. S. Haldane[69], an evolutionary biologist and a founder of population genetics, stated his involvement with God as a scientist:

> *My practice as a scientist is Atheistic. That is to say, when I set up an experiment I assume that no god, angel or devil is going to interfere with its course; and this assumption has been justified by such success as I have achieved in my professional career...*

This is hardly "God has been communicating faithfully," and clearly not for hundreds of years "to the entire human community through the full range of sciences." Dowd's sermon is long, with many sub-sermons included, one of which is detailed in his book and devoted to "Realizing the Miraculous (Virgin Birth and Christ's Resurrection and Ascension into Heaven)". Dowd does not explain how that religious chapter on "Realizing the Miraculous" adds weight to his theotheory on Darwinian evolution. Surely "realizing the miraculous" should have warned scientists that his sermon will have his Christian God in charge of evolution and employ miraculous means. Christians may be happy with this, but scientists should label this chapter as a "God-speak sermon" and quickly move on.

Dowd argues for us to consider more "God-speak" by declaring that all science is "sacred" or "divine" and that natural science theories can be replaced with supernatural "sacred" theotheories. This "marriage" of natural science theories with religion allows Dowd to remove any conflict between natural science and religion that may come up, thereby achieving harmony with science. But such a harmony comes at an unacceptable expense to science, for the scientific community loses control of being able to separate God-speak from natural science.

Assigning "sacred" and "divine" to science concepts should not be done. Hippocrates in the fourth century BC advanced the practice of medicine by removing the "sacred" and "divine" from his medical diagnoses and treatments. Plato later commented on Hippocrates view of sacred diseases:

> *It is thus with regard divine nor more sacred than other diseases, but has a natural cause from the originates like other affections. Men regard its Nature and cause as divine from ignorance and wonder.*

In the case for man's evolution, Darwin and many other scientists have rejected any involvement by a supernatural God. Dowd quotes no research to change that position. It seems as if Dowd overlooked the large majority of scientists who know that they do not need God's involvement to have evolution work. Are many scientists going to accept his Sacred Innovative Design theotheory on evolution and abandon Darwin's Theory? Dowd gives his *I believes* and *I knows*, but his *I knows* lack any proof of God's involvement. For Christian believers, this may be attractive and adequate (they don't demand proof), but for Naturalists the question arises: where are the proofs?

Dowd's book is a "good news" sermon that calls for celebration of how his Christian theotheory improves minds:

> *God is more awesome than ever before, how we can improve our minds now, and how we can appreciate our brain's creation story.*

Naturalists would argue that persons of all religious and non-religious persuasions can improve their minds and appreciate their brain's evolution story without calling for the Christian God to rewrite Darwinian evolution. For scientists the brain was not created by a supernatural God in the biblical sense (instantly), but by Nature's process of natural selection by which the brain evolved as an integral part of our ancestors, including man, over billions of years.

His book (sermon) is for Christian believers and is in disagreement with scientists who argue that man's evolution is a universal process of Nature that works for all—Atheists, Muslims, Hindus, Heathens, as well as Christians. In Christian churches, his subject, *Thank God for Evolution*, would be acceptable as a sermon, but in halls of biblical and secular scholars as well as in public schools and universities, it is not acceptable. Neither God nor Darwin needs the help that Dowd is proposing to give them.

GIFT FROM GOD?

Dowd argues that Nature (evolution) gives gifts to God (religion):

Evolution is a gift to religion, and that meaning-making is a gift to science.

Evolution is a natural process of Mother Nature, and she never gives gifts. She only gives what Nature's tool chest (physics, chemistry, and biology) allows. But why would the supernatural, omnipotent Christian God need a gift from godless Nature? Further, it is baffling that Dowd has God giving "meaning-making" as a gift to science. Man slowly evolved and has done quite well over ninety-eight of the last one hundred thousand years as a Pagan without gifts of "meaning-making." Sometimes science got the meaning-making right and sometimes wrong, but along the way it has developed a process that has been able to correct its wrongs and continued advancing the understanding and meaning of science long before the Christian God appeared on the scene. Further, Dowd overlooks the fact that there are hundreds of gods and religions—Buddhist, Hindu, Zoroastrian, and Islamic—each with its own "meaning-makings" on evolution. Why would any scientist want to wade through an avalanche of hundreds of religious evolutionary "meaning-makings" when he or she has his or her own godless way to make meanings of natural evolution?

To say that a supernatural Christian God uses a natural evolution theory to explain man's creation makes the Christian God superfluous, for Nature has used its evolutionary processes for billions of years without him. It appears that Dowd has demoted God to be a passenger riding on Darwin's back doing none of the hard work of natural selection, but getting the credit for the results. Mixing natural evolution (science) with God's involvement (religion) not only violates the precepts of both science and religion, but also delays serious discussions on how religion and science can independently serve mankind.

THE "GREAT STORY"

Dowd's theme, as he modestly states in his book, is *the great story, the sacred story of everyone and everything*. In his story, the "marriage" of supernatural religion with natural science is pervasive throughout; for Dowd wants to demonstrate that his God has authority over Nature and directs everything.

Down's "Great Story" is the marriage of science and religion seamlessly weaving together science, religion, and the needs of today's world. Because the Great Story and other creation stories of classical religions from native peoples emerged well before the revelation of today's evolutionary cosmos, those venerable stories must be read metaphorically so that the fruits of today's science can stand independently.

Mixing or "marrying" supernatural religion and natural science by "seamlessly weaving together" means that the identifiable tenets and theories of science and those of religion are muddled together to produce Dowd's supernatural Sacred Innovative Design theotheory.

Included is a curious argument for fulfilling the "deep-time potential" of old biblical creation stories written before modern science, or as Dowd says, "ancient cosmologies are creatively reinterpreted to mesh with today's science." Such statements raise questions of why God did not use the "deep-time potential," which he must have known about, and get the "ancient cosmology" right in the first place. One can only presume that this is an attempt to argue, for example, the creation story in Genesis, when creatively reinterpreted is not really scientifically wrong (it is) because it can now be creatively (religiously) interpreted to mesh with today's science. Dowd says:

> *The ancients (Bible) could not have known that God did not aim for the kind of perfection possible through design (Adam and Eve). Rather God favored the slow and rambling paths of emergent evolution.*

Most Christians believe that the ancients who wrote the Bible were inspired by God. Dowd has it a little backward here. It was God telling the ancients about Adam and Eve, not the other way around. How does Dowd know that God did not aim for the kind of perfection possible through instant design? Possibly Dowd forgot that Christians claim the Bible to be God's words, and whether one reads them literally or metaphorically, one would think that God could have gotten his own words right in the first place. Being omnipotent, he must have known what he was doing when he wrote the Bible and gave his account of creating man. There is little credibility for Dowd's God-speak telling us what God favored or what God aimed for. God's biblical words say God created Adam and Eve in a day and not on the rambling paths of "emergent" natural evolution, which took billions of years. Surely God knows the difference between a few days and a few billion years.

Mixing science and religion does produce strange things. In effect, Dowd is arguing that his "Great Story" does not use the old cosmologies of the Holy Bible, for they are just quaint background stories to be rewritten at anybody's will. With his approach of telling God what to do, he proceeds to rewrite religion and science with a "sacralized perspective" that interweaves science and religion, or as he says, "a marriage," which gives us a "holy" understanding of evolution:

> *Believing in "our sacralized evolutionary perspective" and our holy understanding of evolution will.*

Evolution as seen by Darwin and most scientists is a godless theory, not a sacred theory. Nowhere does Darwin (or other evolutionary biologists) ever employ a "sacralized evolutionary perspective." What does a Christian "holy understanding of evolution" mean to scientists working with godless theories? How does one test a theory as to whether it is "sacred" or "holy?" To call it "sacred" or "holy" shifts the understanding of evolution from being a godless theory of Nature testable by science to an untestable supernatural theotheory of God. Such stories may be useful in religious sermons, but not one to be used in a science classroom or laboratory. Science does not use a *"holy understanding"* of anything; just the opposite; it views any explanation of Nature as temporary and works like the devil (a pun) to replace it with a better understanding. Science argues that it has only temporary understandings, never a "sacred" or unchangeable understanding. Science demands testability, and Dowd's supernatural Sacred Innovative Design theotheory of evolution is not testable, for it includes the actions of a supernatural God.

Dowd argues that since God made science and religion, then the mixing of them together will make them compatible:

Science and Religion are two sides of the same coin.

Nothing could be further from the truth. In the natural world science is godless, universal, testable, and impermanent (changeable upon acceptable new information), while religious dictates are supernatural, local to one religion, untestable, and fixed by a God from on high. With no visible compatibility between religion and science, one would argue that they are two different coins, not two comparable sides of the same coin.

Considering the trajectory of science and religion over history, most scientists would part company with Dowd's assertion:

. . . that each (science and religion) is moving in remarkable, previously unthinkable directions.

Just the opposite is true. Science has always moved in directions it never anticipated, for surprises in science are usually how advances are made by observing Mother Nature (physics, chemistry, and biology) do her thing on the ground level. Religion, on the other hand, has been and continues to be sent from on high. Religions with regard to science have steadfastly moved only in one direction—to keep the status quo of religious authority by keeping natural science advances from impacting (conflicting)with it.

The success of science, in spite of religious opposition, has forced religions to change its scope and begin to recognize Nature's authority over natural science here and there. Lightning, for example, was explained by churches as being sent by God. Benjamin Franklin removed God from having any authority over lightning by experimentally demonstrating that lightning was a natural event. Afterwards churches slowly started employing less prayer and installing more lightning rods for the simple reason that Franklin's godless theory of lightning and his lightning rods worked. The use of lightning rods has saved many churches from destruction. There is irony in godless science serving as a shield for God's houses of worship against God's wrath.

Dowd's proposed marriage between science and religion is based on declaring that all of science is "Sacred" or "Divine." In his best preaching style, Dowd declares:

> *I conclude with the claim that a sacred, meaningful view of evolution sanctifies science, realizes religion, and shows that our way into the future, God's will, is obvious and universal. What is this way—perhaps our only way into a glorious future? Christ-centeredness! Evolutionary integrity!*

This mixing of science and religion in the style of an old-fashioned religious revival preacher selling a "glorious future" is charming, but we still need to separate the science from the "god-speak." I think that Darwin would not agree that there is a "sacred view" of evolution or that "evolution sanctifies science."

Darwin's Theory

Dowd does not accept Darwin's Theory, for it does not involve his Christian God in the evolution of man. To have his God involved, Dowd proposes a replacement theotheory, Sacred Innovative Design. But this approach creates a number of problems. First, there is the question of how Dowd talked God into embracing Darwinian evolution, a concept that God never addressed, for in the Genesis story, God only mentions the creation of man and organisms[70] in final form in a day or so, no evolution required. But Dowd no longer likes the biblical creation story which centers a brief encounter with Adam and Eve, so he rejects the Genesis story and uses the godless Darwinian theory of evolution—namely, man evolved over billions of years instead.

Why would God want to have his supernatural narrative, important to so many Christian believers for over two thousand years, changed at this time in order to embrace godless science? Many would argue that God puts salvation over science, but it appears that Dowd argues for science over salvation. Simply, Dowd wants to use the long time frame from Darwin's Theory for the evolutionary period of man. For Dowd, God's instant creation of man is out, and Nature's long evolutionary time frame for man is in. Dowd could have said that the supernatural world of God and the natural world of Nature should be separated; instead he chooses to mix the two to have God in command of both.

By mixing supernatural Genesis with Darwin's Theory, Dowd is forced to confront other tenets of Darwin's Theory, as well as rewriting

Genesis. This brings many problems, for how do you rewrite the Genesis creation story, which happened over a period of only a few days, to reflect Darwin's Theory of man's evolution occurring over billions of years? One approach is to consider Genesis as a metaphorical story, but with this assumption, other problems quickly appear. Does this mean that all of Genesis is to be considered a metaphor, including Adam and Eve and Original Sin? Few Christians would welcome this change.

With a natural science time frame of billions of years for the evolution of man, there cannot be a single Adam and Eve, a single Garden of Eden, or a single Original Sin, but hundreds of thousands. This raises many questions. How does God impart man with Original Sin, an event that happened at only one point in time with one Adam and Eve, when there are hundreds of thousands of Adams and Eves? Dowd's answer is to place God in charge of Darwinian evolution by simply declaring that his supernatural God is omnipotent, has created everything, and has authority over Nature, including Darwinian evolution, and when necessary use a miracle or two. In the Sacred Innovative Design theotheory, God directs the evolutionary process as the creator and designer of all living organisms. A big job: design each cell (there are billions) of every living organism (there are billions) over several billion years so they can evolve according to God's authority. God cannot get help from Darwin's natural selection process, for it is based on chance events (God does not act with chance), so God must do the work of natural selection every minute and soldier on day by day, arranging the DNA, RNA, and the proteins for each of the billions of cells for the billions of living organisms, inserting mutations and copying errors into each DNA for billions of years while selecting the cells that meet God's design criteria for evolution and killing the rest (God selection replacing natural selection). Dowd can only hope that God has the infinite patience to pull this off.

A summary comparison of Dowd's Sacred Innovative Design theo-theory with key tenets of Darwin's Theory is included in the summary of theotheories in Table A.

NATURAL WORLD

Dowd mixes the supernatural world with the natural one throughout his book, and in doing so, his Sacred Innovative Design theotheory is in conflict with many tenets of Darwin's Theory. Dowd attempts to finesse the need for miracles (you can use them or not) by declaring them available, but unimportant to his theotheory:

> *The choice one makes to believe or not to believe, as a literal scientific fact, the miraculous story of. . . is unimportant from an evolutionary religious perspective.*

For the scientific community, the use of a miracle is not unimportant. It is a "deal breaker." Its use places any theory in the natural world into the supernatural theotheory world, removes chance from evolution, and uses God-directed design, which violates natural science.

Dowd sees God's authority everywhere, including in all of science. The beginning of the universe:

> *Seen through sacred eyes, the entire history of the universe can now be honored as the primary revelation of God.*

Again, this is great for a sermon, but most physicists don't have sacred eyes, and working with their natural eyes they cannot see the revelations of the Christian God. As we discussed, Stephen Hawking's view[16] and that of most other physicists is that the universe is godless, and its creation is a result of Nature's processes (physics, chemistry, and biology). An understanding of the universe is being achieved by discovering Nature, one physics theory and one chemistry theory at a time. Its fine for believers through their "sacred eyes" to see revelations of their God, but that is in the supernatural world of God, not the natural world of Darwinian evolution.

CHANCE

Dowd argues to reject chance as part of natural selection:

> *Evolution is not blind chance.*

That is Dowd's religious view, not the view of science incorporated in Darwin's Theory, which does use chance.

DESIGN

Dowd installs God as the innovative designer by directly imposing God on Darwinian evolution.

> *Evolution clearly is the product of intelligent innovation and adaption.*

Clearly proposing "intelligent innovation" is using God as an intelligent designer. This is simply a replay of the old Intelligent Design theotheory, which most scientists (including the other Christian authors reviewed) have rejected. Darwin's Theory argues that the design of man is without a designer, for it is designed by Nature's process with random variation and natural selection.

COMMON ANCESTOR

Dowd then tells God how he got it wrong in Genesis and that God really created Adam and Eve by natural godless evolution. Dowd does not seem uncomfortable telling God what he did in spite of what God said (Genesis):

> *This {natural godless evolution} is how our God, the Creator,*
> *made Adam and Eve and the rest of us.*

This is a religious statement; Dowd cannot stop telling God what to do—in this case "God, you got it all wrong with your words in Genesis." If natural godless evolution is how God made Adam and Eve and the rest of us, then the book of Genesis must be interpreted differently, for natural evolution does not have one Adam and Eve instantly appearing fully formed in a Garden of Eden discussing life and the philosophy of good and evil with a talking serpent.

Instead, natural evolution has given us an evolving design and a moral sense without a God by Nature's evolutionary processes. Our ancestors range from green pond scum to multi-cell bacteria, to fish, to animal, and through a hundred million Adam and Eves eventually to modern man.

Evolution never stops; there is no final design or image for all life forms, including man. The fossil record documents the changing images of our ancestors, which have included the great ape eight million years ago before our branch of the tree split and evolved to chimps four million years ago and later split into the branch of Homo sapiens in the last one hundred thousand years.

MORAL SENSE

"Original Sin" is a singular biblical event occurring in the Garden of Eden with Adam, Eve, God, and a talking serpent. This is a theological event described in the Christian narrative. In our natural world, man's moral sense evolved as our ancestor's brains evolved over hundreds of millions of years through millions of Adams and Eves. As a product of evolution, man is born with an evolved moral sense and does not have the supernatural burden of his forefathers having committed a religiously sinful act—Original Sin.

By putting God as the designer Dowd must invent a theotheory that explains how "Original Sin" was given to man by having God hijack natural evolution and plant "seeds" of the supernatural "Original Sin" into the brains of our ancestors.

> *Even the most innocent babe carries the seeds of "Original Sin" in its brain stem and limbic system.*

> *Such "Original Sin" having been imprinted by evolution in our brain is the cause of our sins of today.*

Dowd assumes that every child ("even the most innocent babe") in the evolutionary tree carries the Christian precept of Original Sin. As discussed, the opposite appears to be true, for evolution has given "even the most innocent babe" an evolved moral sense.

Further, the concept of Original Sin doesn't pass the test of natural evolution, for there could not have been an original pair (Adam and Eve) to disobey God's commands in the first place. Francisco Ayala notes:

> *There is no known mechanism by which the human species might have arisen by a single step in one or two individuals only, from whom the rest of mankind would have descended.*

Without an Adam and Eve, there could not have been an "Original Sin." What about other "babes" born to those of other religions who do not subscribe to Christian precepts? This is a religious argument, an invention by Dowd, to keep the supernatural concept of Original Sin alive today when it should have been rejected along with the concept of a Garden of Eden after we learned that evolution took place over billions of years, not in a few days.

Our brain's neural network does include shards from our early ancestors' brains, including instincts from fish, reptiles, monkeys, and apes. After millions of years of brain evolution, our neural net-work has grown to be an amalgam of many old "seeds" embedded in a larger neural network.

Dowd does not hide the problem that Christianity has with mor-als, and indeed, he relays some of the horrors described by biblical descriptions of killings and slaughters of women and children by Hebrew armies in the name of their God. He notes:

> *Of course the image of God portrayed in Scripture is sometimes terrifying.*

These immoral acts of killing the populations of whole cities on a grand scale rightly point out horrors in the scriptures that are a major source of resistance by people to Christianity. How can one want to be a Christian when their history is filled with immoral acts? His defense is to replace God with Nature's evolution for cover.

> *Evolution offers a much less vindictive and far more venerable understanding of God {than} the one portrayed in the Bible and the Qur'an. This should not be surprising, nor is it a denigration of Scripture. In a divinely emergent Cosmos, how could it have been otherwise?*

One would not have expected a Christian preacher to turn to the Agnostic Darwin for the rescue of Christian morals. Dowd proposes that Darwinian evolution has given us:

> *. . . much less vindictive and far more venerable understanding of God {than} the one portrayed in the Bible.*

The killings in the Bible are God's words, and Dowd wants us to use Darwin's words instead. For a minister to use Darwin instead of the Bible for an understanding of God's words is a most telling statement. Does Dowd know he is giving Nature the authority over man's creation-evolution?

When using "divine" or "sacred" as an adjective modifying natural science, it is easy to fool yourself. Science does not have a religious side; it is without religion. Nature does not reflect a divine anything; it is godless. The gambit Dowd uses is to borrow whatever concepts he likes from Darwin, put the label "divine" or "sacred" on it, and tell us that God has approved it. Dowd argues that we should fool ourselves.

Christianity is more significant and deeper than the picture painted by Dowd and does not need to ride on the back of Darwin to be relevant or sacred. Dowd is far too busy trying to rewrite parts of the Bible by inserting Darwin rather than understand the deeper meaning of the Christian narrative held by millions of Christians. God's words, his narrative given in the Bible, are important as they stand and do not deserve to be rewritten by Dowd.

AGREEMENTS

Removed from his discussions on evolution are some positive observations to be found in Dowd's book. One is that both religion and science make sound contributions to mankind and will continue. Dowd correctly observes that

> . . . neither religion nor science will drive the other into extinction.

I agree with this conclusion but differ in that science has never tried to drive religion anywhere. The "driving out" rhetoric is a throwback to the early attempts by Christians after Darwin to do just that. It was William Jennings Bryan in the 1920s who said:

> We will drive Darwinism from the schools.[71]

Dowd does make some assertions that are not helpful to resolving conflicts, with strident words that:

> *. . . science must triumph over supernatural religion and render it ineffectual if our species is to survive.*

It is arguments like "science must triumph" that keep religious congregations in a combative mode, and some ministers use these arguments to frame discussions of evolution and science as being in a death struggle, a war. Many labels given to non-Christians, such as heathen, secular, materialist, and Atheist are pejorative, and Christians are asked to defeat them and their amoral "secular materialism" before they overrun Christianity and the country. Such arguments must be abandoned, for our thrust should be not on how to win, but on how to respect the independence of each.

On another note of agreement, Dowd has observed that many changes have occurred in science and religion and that they should not be feared.

> *Change is to be welcomed, not feared.*

His call to welcome change is reading from the same page as science, which thrives on change. This leaves a positive note that religion and science may continue to learn from changes. However, history leads to caution here, for if the change is to be the one advocated by Dowd—mix religion and science together—it should be feared. If change is within religion or within science, it is positive and should be welcomed. Whatever the change may be, the independence of science and religion must be honored. One observation about scriptures is also on the mark. Becoming scripture means that the bureaucracy has taken over and changes in religious (or any other) dogma are no longer welcomed; only the status quo is welcome.

When a story becomes Scripture, it ceases to evolve.

Understanding that scriptures get frozen in time and that change is not to be feared, one must ask why Dowd works so hard to have several old scriptural passages (Adam and Eve, Original Sin, etc.) incorporated and modified into his Sacred Innovative Design theotheory.

SUMMARY

There is much to like in this sermon by a Christian minister who openly discusses some of the shortsightedness of the Christian narrative with regard to science. He is a Christian minister-writer who promises a happy marriage with science for the future. Dowd gives a spirited discussion of science, albeit incorrect in many places. His book may expose some Christians to a new science concepts they would not have otherwise heard. So what is the concern? Little, if Dowd's book is understood as a religious sermon leading Christians not to fear science and to embrace it for its value independent from science.

There is much to be concerned about when examining the details of his proposed marriage of religion with science, for Dowd violates many of the tenets of Darwin's Theory, accepts that miracles may be used, rejects chance as part of evolution, and argues for an Innovative Designer (God) while rejecting a common ancestor for all life. These are not science statements, but pieces of Dowd's religious theotheory of evolution (Sacred Innovative Design). Naturalists find Dowd's God-centered theotheory an unacceptable replacement for Darwin's Theory, which is Nature-centered.

Arguing that theotheories are useful to explain evolution in the natural world violates natural science theories. Redefining science to be "divine" or "sacred" mixes in his religion and produces a theotheory, which harms both religion and science. How many public university biology departments would want to teach *Sacred Innovative Design* in their biology departments? Can you picture a biology professor giving his class an assignment to devise an experiment to differentiate between a secular and a "sacred" evolutionary process, or to explore for "divine" fossils? The words "sacred," "holy" and/or "divine" have no meaning in science, for by using these adjectives in natural science, it is easy to fool yourself. The author Salman Rushdie, who was the object of a "sacred" death threat, a Muslin fatwa, rightfully argues that "sacred" should not be viewed as a positive thought:

> *The idea of the sacred is quite simply one of the most conservative notions in any culture, because it seeks to turn other ideas—uncertainty, progress, change—into crimes.*[73]

Christianity has more to offer than using a God as a replacement for Nature in Darwin's Theory. To scientists, a fatal flaw in Dowd's sermon—adding the religious adjective "sacred" to natural science theories—marks any science theory as a theotheory, a religious one, and must be avoided outside of the church. This book should not be a welcome addition to a science classroom.

But Dowd claims otherwise, for he proposes that the marriage of religion and science will bring harmony between science and religion. Christians may rejoice with Dowd telling them what God's "good news" is all about, but I think biblical scholars will not. Most scientists rebel against attempts to "sanctify" science or to follow a religious ("Christ-centeredness") way in science. But Dowd sees scientists rejoicing when being told how the "sacred" science evolution theotheory "sanctifies science."

It is an interesting sermon for Christians and ends with Dowd's prediction (given in his subtitle to the book): *"How the marriage of science and religion will transform your life and our world."* However, non-believers will not see their life transformed by the religious arguments Dowd uses. As a good Christian minister, Dowd ends his sermon by asking his readers (congregation) to "testify and share the good news" about mixing religion and science. This is a fitting end for this sermon to Christians. For non-Christians, it is another example of a theotheory (called Sacred Innovative Design) involving science and natural evolution, which needs to be carefully kept in the supernatural category.

The book's title, *Thank God for Evolution*, tells us that his supernatural God has been given credit for Darwin's Theory of evolution—a clear case of a natural science theory being hijacked by religion and mis-applied to support the supernatural Christian God's authority over Nature. In ending Dowd says:

> *To agnostics, humanists, Atheists, and free thinkers I promise that you will find nothing here that you cannot whole heartedly embrace as being grounded in a rationally sound, mainstream scientific understanding of the universe.*

There is not much here that I can embrace. The unlimited and unabashed mixing (a marriage) of religion with science is a disservice to religion and science.

Dowd's profession is that of a minister whose duty is to propagate the Christian faith, and his book reflects his zeal to do this, for it is indeed a sermon about bringing "the good news" of religion to all. With this sermon he obviously believes that by saying *thank God for evolution*, he has achieved his ministerial goal of declaring that all is in harmony between religion and science.

However, Naturalists see no harmony, do not like to see Darwin's Theory mutilated, and are convinced that God does not have the authority over Nature or evolution.

The Language of God-Collins

Francis Collins (*The Language of God: A Scientist Presents Evidence for Belief*) is a respected scientist, government employee, and a self-declared Evangelical Christian. His belief journey from an Atheist to an Evangelical is an inspiring story for Christians and is briefly summarized in his book. He discusses how he reconciles his faith with scientific knowledge, thus removing potential conflicts presented by Darwinian evolution with his religion.

Collins feels compelled as a Christian to place Nature in a subservient position to his God by declaring that God "established natural laws," of the universe, specifically:

> *God, who is not limited in space or time, created the universe and established natural laws that govern it.*

Immediately this places God over Nature. Having God make the natural laws, Collins then singles out Nature's Darwinian evolution as a natural theory that works:

> *Evolution as a mechanism can be and must be true.*

With these two statements, Collins defines his dilemma: he wants his Christian God to have authority over Nature's theories (God established the natural laws), while wanting to use Darwin's godless theory to describe man's evolution, for it works. But from his research, he knows that the mechanics of Darwin's Theory, that is, natural selection works by the use of chance to operate on the random DNA mutations. But evolution driven by chance presents a direct conflict with his belief that God directs evolution according to "God's plan."

To overcome this conflict, Collins modifies Darwin's Theory with an invented workaround replacement theotheory, called Theistic Evolution, which removes chance and inserts God as the designer who works with certainty. But the mixing of science and religion together in his theotheory conflicts with the fundamentals of both science and religion. Natural science theories are no longer science if the supernatural is included. Religious theotheories do not need scientific proof, so his Theistic Evolution theotheory is fine for sermons, but bad for science lectures.

Collins has experienced scientific research and the thrill of discovery from his genetic research. His book's title states his belief that DNA is *The Language of God*. Collins may wish to believe, but his subtitle, *A Scientist Presents Evidence for Belief,* promises evidence, but none is to be found in his book.

What we do know is that DNA, the language of life, was discovered by James Watson and Francis Crick and that it has worked within living cells for over three billion years, from the time of our first common ancestor to us, without any help from God. DNA research has been accomplished by Atheists, Buddhists, Hindus, Zoroastrians, and Christians and has provided detailed support for Darwin's Theory of evolution.

The DNA's molecular language is universal among all living organisms, as the research by Collins and others have clearly shown. No evidence is presented as to how the Christian God interacts (encodes and decodes) with DNA or that DNA is different from that which godless Nature has been using for billions of years as the language of life.

If Christians or believers of other religions want to claim it and invent a religious theotheory to use, they are welcome to do, but one would think that they should, at least, be polite and attach a little note saying, "Thanks, Mother Nature, we Christians are borrowing your beautiful, godless DNA molecule for use in a supernatural story we are writing to praise our God's handiwork." Thus, the conflict is enjoined; Who do we thank for the DNA molecule and its role as the language of life describing evolution, God or Nature?

THEOTHEORIES

Although several theotheories have been invented and tried before, Collins states that the Theistic Evolution theotheory he embraces is the most scientifically consistent and spiritually satisfying of the Christian evolutionary theotheories that have been put forward so far. Previous theotheories, such as Creationism and Intelligent Design, are, in his opinion, inconsistent with today's science, a position supported by most scientists. Collins' objections to Scientific Creationism are its failure to reject the biblical flood from scriptures. For the Intelligent Design theotheory, Collins' objections are:

1. Evolution promotes an Atheistic worldview and, therefore, must be resisted by believers.
2. Evolution is fundamentally flawed, since it cannot account for the intricate complexity of Nature.

3. If evolution cannot explain irreducible complexity, then there must have been an intelligent designer involved somehow who stepped in to provide the necessary components during the course of evolution.

His objections to ID are consistent with those of the scientific community. Of particular note are the examples of research, which have demonstrated evolution with great biological complexity, such as blood clotting in Nature.

Although aware of the rejection of previous theotheories, Collins nevertheless proposes using another, Theistic Evolution, which has also been proposed by other Christians, including the Catholics. Collins has summarized his supporting view of Theistic Evolution with six premises listed in his book and later statements at a lecture at UC Berkeley. These premises, if given as a statement of religious faith, present no conflict, for religions are allowed to view natural science from their religious perspective. But Collins presents these from a scientific viewpoint.

Collins' key premises for Theistic Evolution are listed below. The parts of a premise that conflict with natural science are given in italics, and comments are added to explain the conflict.

Premise 1. The universe came into being out of nothingness approximately 13.7 billion years ago.

Comment: Agrees with natural science.

Premise 2. *Despite massive improbabilities, the properties of the universe appear to have been precisely tuned for life. Almighty God, who is not limited in space or time, created a universe 13.7 billion years ago with its parameters precisely tuned to allow the development of complexity over long periods of time.*

Comment: Collins' Theistic Evolution theotheory installs God as the creator of the universe and designs ("tunes") its parameters to have man evolve. Requiring a precisely tuned universe is the old argument for a designer God (Intelligent Designer) who has the authority over Nature and would create and design the universe to support man. Naturalists argue that the universe was created by Nature, created without a creator and designed without as designer. What we observe is that the universe was created following the laws of Nature (physics, chemistry, and biology). The panoply of cosmic objects, galaxies, stars, etc. and their movements, as well as our planet spinning in its orbit around the sun only obey natural forces, no God required.

Premise 3. While the precise mechanism of the origin of life on earth remains unknown, once life arose, the process of evolution and natural selection permitted the development of biological diversity and complexity over very long periods of time.

Comment—This is the position of science, so there is no conflict. But Collins takes an additional step in the next premise by giving God credit for Nature's evolution.

Premise 4. *God's plan included the mechanism of evolution to create the marvelous diversity of living things on our planet. Most especially, that creative plan included human beings.*

Comment: Requiring God to have a plan for life (mechanism of evolution) is but more of the intelligent designer argument. This requirement is, of course, in conflict with Darwin's Theory, where chance (random variations and natural selection) places a critical role. By removing chance and substituting God to direct evolution, Collins has moved the Theistic Evolution theotheory out of the natural world and into the supernatural world of religion.

Premise 5. Once evolution got underway, no special supernatural intervention was required.

Comment: No problem for science, but big problem for Collins, for evolution uses natural selection, which is driven by chance that cannot follow God's certain plan.

Premise 6a. God gifted humanity with the knowledge of good and evil (the moral law).

Comment: Having God gift man with moral laws requires supernatural intervention into man and/or his ancestors during evolution. That requires a miracle. This act defies natural selection, which provides an evolved moral sense in man without the intervention of a God.

Premise 6b. Humans are part of this process, sharing a common ancestor with the apes.

Comment: No problem with Darwin here.

Premise 7. But humans are also unique in ways that defy evolutionary explanations and point to our spiritual nature. God has given man the Moral Law (the knowledge of right and wrong) and the search for God that characterizes all human cultures throughout history.

Comment: Collins steps on science and Darwin's toes here. All humans are unique at the DNA level, but then he violates Darwin's Theory by declaring that man is *unique in ways that defy evolutionary explanation.* If Collins' theotheory, Theistic Evolution, defies evolutionary expectations, then Collins must tell us what does the defying. Is it God using a miracle to insert the moral law? Where is the proof of the modification of Darwin's Theory? But Collins gives no proof to support his Theistic Evolution theotheory.

There is a conflict with science when Collins says God gave man the moral law. From natural evolution, man is born with an inherited moral sense, which is expanded and refined subsequently by experience and education. We are not born with a specific religion's set of morals. Later, other morals from Christians, as well as those of all other religions—Hindus, Zoroastrian, etc.—may be exposed to children.

DARWIN'S THEORY

Collins does not accept Darwin's Theory as defined by the scientific community, so he is forced to propose a replacement, the Theistic Evolution theotheory, which requires modifications to Darwin's. The key tenets of Darwin's Theory of natural selection are modified or rejected by Collins are outlined below.

NATURAL WORLD

Darwin's Theory operates in the natural world. The Theistic Evolution theotheory Collins proposes operates in a supernatural world in which miracles are acceptable. Collins tries to explain when God may use miracles:

> *Perhaps on rare occasions God does perform miracles. But for the most part they are rare.*

That's fine for a religious theotheory in God's supernatural world, but miracles are not scientifically acceptable for any science theory in the natural world. It does not matter whether they are rare or not, they are not allowed. Collins is aware of the sensitivity of coupling miracles and science and expends considerable effort discussing their use in Theistic Evolution. Collins turns to fellow Christian and scientist John Polkinghorne[74] for support:

> *Miracles are not to be interpreted as divine acts against the laws of Nature (for those laws are themselves expressions of God's will) but as more profound revelations of the character of the divine relationship to creation. To be credible, miracles must convey a deeper understanding than could have been obtained without them.*

If we apply a rational meaning of a miracle, an absence of knowledge, to the above paragraph, we can see the full non-sensible meaning of this paragraph by Polkinghorne:

To be credible, {the absences of knowledge} must convey a deeper understanding than could have been obtained without {the absences of knowledge}.

Farther along in his book, Collins reiterates the need for miracles:

I believe that possibility (of supernatural miracles) exists, but at the same time the "prior" {possibilities} should generally be very low.

Science forbids any miracle, even if it is a just a very little one. Collins has hopelessly jumbled God's supernatural miracles with Nature and any distinction between the natural and supernatural worlds. The fact that a scientist resorts to a theotheory (Theistic Evolution), which requires supernatural miracles to infuse certain attributes from God into man only affirms that Collins has given the authority over Nature to God. The abandonment of support for the scientific base of Darwin's Theory makes his book a religious book, which Christians may find interesting, but a book that should not be used in science classes.

Chance

The title of Collins' book, *The Language of God*, refers to the DNA molecule providing the language to instruct a living cell to be assembled, to grow, to function, and to die. As a language by which God communicates with life, DNA would be the carrier of messages which can be changed by chance from random mutations and copying errors. His fellow biologist and Christian, Ayala, sees chance as an essential part of evolution:

Chance is, nevertheless, an integral part of the evolutionary process. Mutations that yield the hereditary variations available to natural selection arise at random.

Specifically, Ayala is saying that Darwin's Theory of evolution works quite well with chance, in fact, it needs chance. But Collins knows that he must remove chance from evolution in order to have his God in charge. To get out of this dilemma, Collins appeals to a miracle—change how time works:

> *Evolution could appear to us to be driven by chance, but from God's perspective the outcome would be entirely specified. Thus God could be completely and intimately involved in the creation of all species, while from our perspective, limited as it is by the tyranny of linear time, this would appear a random and undirected process.*

Tinkering with the scientific definition of time by saying that time in the supernatural world (God's perspective) is different from the time in the natural universe may be an interesting theotheory in the religious world, but it is an unsupportable religious view in Nature's scientific world. If Collins has any evidence on the physical change of time, he should share it with the scientific community. After all, Einstein fiddled with time and space and gave us the Theory of Relativity, but then, he did not look to miracles for help.

DESIGN

Collins makes God the designer (to follow God's plan) in evolution:

> *God's plan included the mechanism of evolution to create the marvelous diversity of living things on our planet. Most especially, that creative plan included human beings.*

With God as the designer (God's plan) in evolution, there is conflict with Darwin's most important idea—Nature through evolution creates designs without a designer. But Collins refuses to accept

Darwin's evolutionary explanation for the design of life and replaces it with God as the designer in his supernatural Theistic Evolution theotheory.

The DNA language of life refers to all life, viruses to vertebrates, which includes not only normal life forms, but also misshapen life forms. Mutant DNA instructions can result in a child with two heads, malformed limbs, and all of the other occurring human abnormalities witnessed in the world. If God is the DNA designer following "God's plan," then he is responsible for all, the good and the bad. Collins does not give an answer to why his God, the architect of the DNA, would send the polio virus with instructions to search predominately for children to infect, to maim and to kill. It is difficult for Christians to acknowledge that such killer messages sent by DNA are communications following God's plan.

Common Ancestors

Collins is well aware that all living organisms have a common ancestor and that we have a DNA trail from our earliest ancestor to us that follows evolutionary expectations. But Collins finds it necessary to violate this natural DNA trail by proposing that:

> God created humans unique in ways that defy evolutionary explanations and point to our spiritual nature.

Tests of DNA from samples in our long line of ancestors so far have not supported this proposal. Nothing that is evolutionarily unique has been found in the human DNA—in fact, just the opposite.

Moral Sense

The appeal of Christian moral laws was an important factor ("a signpost to God") that attracted Collins and led to his conversion to

Christianity. And indeed there are Christian moral laws that appeal to many, and they should be held in respect. Collins' discussion of morals should have ended here, but he takes the moral laws he admires out of his belief world and attempts to explain how God gave man his moral laws to man. Collins argues that God inserts his morals into man and/or his ancestors at a specific time during the evolutionary process:

> *After evolution had prepared a sufficiently advanced "house" (the human brain), God gifted humanity with the knowledge of good and evil, with free will, and with an immortal soul.*

This is in conflict with natural evolution, which does not have a line dividing the evolution of the morally non-advanced from the morally advanced human brain. Nature sees the evolution of the brain as a continuous process of the development of the vastly complicated neural network. There is no one point along the continuum of evolution to define a "sufficiently advanced house" that the evolution of the many factors involved in man's moral sense had not already started on its evolutionary path. Instead of accepting the natural, gradual evolution of our moral sense, Collins is trying to inject God's morals *in toto* at one point in the process.

This argument goes far beyond the scriptural creation of man and burdens God with accomplishing the task of inserting knowledge of moral laws (he does not say which ones) at a specific point in the evolution of the brain. Collins refuses to agree with the biological evolution of human morals:

> *If the moral law is just a side effect of evolution, then there is no such thing as good or evil.*

But in Nature the evolution of our moral sense, as well as other attributes of the brain, is integral to man's evolving mental capabilities. Man's social and moral senses are part of man's evolution and what makes us human. The whole package evolves together. Morals are not a side effect of the biology of evolution; they are an integral product of evolution. Patricia Churchland addresses Collins' comment:

> *Actual human moral behavior, in all of its glory and complexity, cannot be cheapened by the false dilemma: either God secures the moral law or morality is an illusion.*

Collins claims that the Adam and Eve story is a powerful allegory of God's plan for the entrance of the spiritual nature (the soul) and the Christian Moral Law into man. He then relies on his faith to argue that man's sinful nature is a result of the Original Sin from the actions of Adam and Eve.

> *We humans used our free will to break the moral law, leading to our estrangement from God. For Christians, Jesus is the solution to that estrangement.*

Collins declares that man's soul was directly created by God.

> *God gifted humanity with an immortal soul.*

The Bible doesn't describe the soul in detail, except that it is immortal. The understanding of what is a soul is not consistent among Christian denominations, much less among the thousands of other religions. Collins does not describe what he envisions the soul to be or how it would be gifted to man.

Is Collins' view of the soul the same as that of the Catholics, which describes it as a separate living entity in the body? Referencing immortality in a book discussing evolutionary biology defines the argument as a religious one. A supernatural God is free to gift "an immortal soul" to whomever he wishes, for natural science has nothing to say about either God or immortality. Certainly Darwin had nothing to say about immortality.

Neuroscience views the "soul" of a person as something far different. If the "soul" is described by science, it would be likened to the output of our brain with all of its integrated calculational and memory complexity, none of which are not gifts from God. They are a result of man's evolution, so science can only conclude that the "soul" is but the natural output of the brain.

Finally, Collins does not use the advances made in neurobiology explaining the evolution of morals as a natural result and not divine gifts from any god. He notes his resistance to natural science's explanation:

> Others will object that the Moral Law is simply a consequence of evolutionary pressures. This objection arises from the new field of sociobiology, and attempts to provide explanations for altruistic behavior on the basis of its positive value in Darwinian selection. If this argument could be shown to hold up, the interpretation of many of the requirements of the Moral Law as a signpost to God would be potentially in trouble.

Collins' view that Christian morals are a "signpost to God" is in trouble from the scientific understandings coming out of the Neuronian Revolution. The evolution of a moral sense is a result of and an integral part of Darwinian evolution.

SUMMARY

Collins' approach to forging a reconciliation of his religion with science is to mix science and religion and invent a theotheory, Theistic Evolution, which combines them, and give his Christian God the authority over Nature's science. His book title, *The Language of God*, declares that DNA is a language used by God. He supports this view by proposing his Theistic Evolution theotheory as a replacement for Darwin's Theory which employs the Christian supernatural God for the design of man. The use of God, who is always certain, allows Collins to remove chance from the design. Collins gives God the job of being the designer and general contractor of the DNA in living organisms.

This is contrary to Darwinian evolution, which unfolds in the natural world by Nature's process of natural selection, which is driven by chance. Removing chance erodes the scientific rigor (and beauty) of Darwin's Theory.

Collins rightfully acknowledges that the truths of his Theistic Evolution theotheory can be tested only by the spiritual logic of the heart, the mind, and the soul; an admission that his book is not about science, but about a religious description of evolution. If he had labeled his book *The Language of God: A Religious View* and declared it a sermon, there would be no arguments, for believers can discuss science as they see fit. However, his book is addressed to the general population and exposes his approach of mixing science and religion together to the public; hence, this critique.

Collins argues for supporting the Theistic Evolution theotheory, for he believes it will bring a "truce" between religion and science:

> *A truce in the escalating war between science and spirit. The war was never necessary. Science is not threatened by God; it is enhanced. God is most certainly not threatened by science; He made it all possible.*

Unfortunately, the conflicts, or "war," have been necessary for science to advance, for it had to remove many dogmas imposed by the

Christian church. Collins only continues the "war," for he insists that Nature and science are subservient to his God.

Collins' religious viewpoint—"He (God) made it (Nature) all possible"—only appears when viewing through his Christian lens, a view that fails at times to see the workings of Nature. It is disappointing that Collins does not acknowledge Darwin's greatest contribution to the understanding of man's evolution—man does not need a designer God, for with Nature man is designed without a designer.

Finally, maybe from frustration, Collins feels the need to attack Atheists:

> *It is hoped that the arguments presented within this book will convince you that of all the possible worldviews, atheism is the least rational.*

This hope does not agree with the facts about the worldviews of scientists in their own field of evolutionary biology. A worldwide survey[75] of his fellow senior evolutionary biologists showed that a majority (87 percent) said that they had no religion. This group seems to have written many rational scientific papers. Collins attacking Atheists is hardly a path to forging a truce and finding harmony among his fellow scientists.

Christians will find Collins' views useful to learn how an eminent scientist, who is also a Christian, personally comes to grips with the difficult task he gives himself—finding a mixture of Christian supernatural precepts with Darwin's Theory, which will give harmony. Unfortunately, Collins' solution, the Theistic Evolution theotheory to replace Darwin's Theory, is but one more Christian theotheory, which, like the others before it, fails the tests of natural science and, as such, is rejected by his peers in the scientific community. But there are many Christian believers who do accept it.

BioLogos Website

After writing his book and leaving the government (for a short interval), Francis Collins established the BioLogos Foundation Web site in order:

> . . . to address the escalating culture war between science and faith in the United States.

The conflicts between religion, science, and the state are not part of a "cultural war," but a power struggle over who has the authority over man's evolution—God or Nature? It is basically a religious war on natural science fought by Christian writers who insist that their God has the authority over Nature's evolutionary processes. Science is involved only because religion has invaded its turf and is attempting to change Nature's theories, such as Darwin's, by either trying to have the state throw it out of public schools, or modifying it. If science were given its independence from religion, there would be no "war" for natural science would have the authority over science theories in the natural world. This would allow each to do their best; Christians to do their thing in the supernatural world of God and science to do its thing independently in Nature's world.

The BioLogos Foundation announced that it was established to explore, promote, and celebrate the integration of science and Christian faith, but it appears that Christians gathered at the BioLogos Web site wish to continue to fight Collins' "war."

> BioLogos is led by a team of scientists who believe in God and are committed to promoting a perspective of the origins of life that is both theologically and scientifically sound.

Three authors of the books reviewed, Collins, Falk, and Giberson, are members of this team, and as would be expected, their messages basically reflect the same point of view—to promote "a perspective that evolution can be both theologically and scientifically sound." This is, of course, an impossible task, but they continue to try.

> *BioLogos represents the harmony of science and faith. It addresses the central themes of science and religion and emphasizes the compatibility of Christian faith with scientific discoveries about the origins of the universe and life.*

The concept they miss is that faith and science on the origins of life are incompatible in the natural world. It is the muddling of the two worlds together that makes them incompatible. The Christian faith can hold whatever views they choose in God's supernatural world, but these views cannot be brought into the natural world without conflict. Each view can be important in its world; it is the mixing of the supernatural with the natural that makes them incompatible. Unfortunately, the search for harmony and compatibility between science and religion associated with this Web site obscures what should be done in the real world: understand the differences[76] and build respect for the independence of religion and science. The "feel good" fictions of "harmony" and "compatibility" need to give way to a productive relationship where each has a respected independence.

CHRISTIANS AND THE STATE

Many government employees who are Christians have come before Collins. One notable was President John Kennedy, who was the first Roman Catholic to win the presidency. He faced a situation when running for president as a Catholic in a country that had many secular laws in conflict with Catholic laws (abortion, divorce, etc.). Only Protestants had been elected president before, so there was an intense spotlight on Kennedy and how he would place his religion in reference to the secular laws of the country. Kennedy's public answer was: "I believe in an America where the separation of church and state is absolute." He declared that he would not mix the two together, and in matters of government, the Constitution would always be his guide. His private life, including his religion, was to be his own. Kennedy was elected, and his service as president was never marred by mixing or subjugating democracy or science to his religion. He remained a Catholic.

Collins, a professional scientist in the government, has, like Kennedy, placed his governmental obligations, rules, and laws above those of his religion, and like Kennedy, he has honored his profession (government employee) while remaining a Christian. However, Collins' actions supporting his religion are not as clear cut as those of Kennedy. His book argues for a religiously based theotheory that conflicts with secular science, and his book is not suitable for the country's secular public classrooms. Further, in one chapter of his book, "An Exhortation to Scientists," he lists details from the biblical Genesis and attempts to sell his religion to the general public by suggesting that fellow scientists should follow his Christian beliefs in order to find happiness. This is a disregard for the separation of church and state policy followed by Kennedy, Thomas Jefferson, James Madison, and other former presidents. One cannot envision any one of these three presidents writing, selling, and actively promoting a religion-based book based on their beliefs that could not be used in a public school science classroom.

Coming to Peace with Science-Falk

D arrel Falk, a biologist, a college professor, and an Evangelical
Christian, explains in his book *Coming to Peace with Science:
Bridging the Worlds between Faith and Biology* how he personally has
been able to come to peace with modern biological evolution and
retain his Evangelical Christian beliefs. Falk is also a member of
the BioLogos Foundation, which advocates for the compatibility of
Christian faith with scientific discoveries on the origins of the universe and life.

His book is a sincere effort to share his Christian views on evolution with his students, church, and other Evangelicals. He uses the
metaphor of "building a bridge" between his faith and natural biology to explain how he overcame conflicts. Unfortunately, the bridge
metaphor is a poor choice for discussing conflicts with science. It may
have been useful for him personally in religious studies, but as a process, it fails when it is used to connect religion to the natural world
of science.

His book addresses biological evolution and the Christian
Scriptures and is written for an audience of Evangelical Christian

students who are taking their first steps away from a Christian high school science education where the norm was a literal reading of the Bible's Genesis creation. Falk approaches this as a Christian talking to other Christians and attempts to soften some of the positions of Darwinian evolution that students face when introduced to college-level evolution classes. In his discussion of evolution, Falk clearly says straightaway that he is wearing "faith spectacles:" when discussing science from a supernatural viewpoint—that is, from Christian theo-theories on evolution, not from a science standpoint. His task is to bring harmony to the Evangelical students from the conflicts presented by Darwinian evolution.

Falk's book gives an appreciation to the conflicts a young Christian student faces in attempting "Coming to Peace with Science." First he believes that he cannot use the normal words used by science when discussing evolution; for example, Darwin's Theory of natural selection is replaced by Gradual Creation, a scientifically incorrect replacement. Additionally, reference to Darwin's Theory are omitted. Second, creation is not interchangeable word with evolution.

Unfortunately, in setting the stage for his discussion on Darwinian evolution, Falk places his students on a false path by quoting Phillip Johnson, a lawyer and a proponent of the failed Intelligent Design theotheory. Johnson argues that statements by proponents of natural science are anti-theistic:

> *The literature of Darwinism is full of anti-theistic conclusions, such as that the universe was not designed and has no purpose, and that we humans are the products of blind natural processes that care nothing about us.*

We must pause here and correct the misconception that Johnson's words foster, that is science is anti-theistic. This is an inaccurate, time-wasting diversion. Natural science addresses man's biological evolution without any consideration of a God. Science is not anti-theistic; it just does not address God's supernatural world, for God is not needed for science in the natural world. Darwinian evolution's natural selection is a natural process that employs only Nature's tool chest. It is a blind process, for it employs chance and does not employ values, such as caring about the morals of its results. Evolutionary biologists say this directly—for example, Richard Dawkins has written extensively on Darwinian evolution and summarizes:

> *We humans are products of blind natural processes that care nothing about us. Our design is without a designer.*

This statement is in the natural world, so it does not conflict with the Christian supernatural view of biblical creation of the universe and man. It is not anti-theistic; it is just what science sees in the natural world. Such statements may sound stringent to Johnson and Falk, but they are the same as those used by the scientific community.

Johnson's quote reveals that he just doesn't get it—that is, understand natural science. Nature is not constrained about "caring about us." It is one of the strengths of natural science that it is not judgmental about caring or giving values to man, beast, or virus. It is not anti-anything; it is about scientific processes using Nature's tools (physics, chemistry, and biology) doing Nature's work in which a God is not used. The theory of gravity is not anti-theistic because it does not require a God. When you drop a hammer and gravity pulls the hammer down and hits your toes, you can complain about your error in dropping the hammer, but not about gravity being uncaring; it was just doing its job. Gravity is a blind natural force that cares nothing about your toes.

Falk then says it is sad that non-Christians fail to see the meaning of the Christian creation story. This, of course, may be true for non-believers who cannot see what Christian believers see, for they have no "faith spectacles." The Christian belief in a supernatural God gives them a view of the supernatural world and blessings that non-Christians do not have. But non-Christians are not sad and readily grant Christians (and those of other faiths) their religious views and the blessings they may bring, for they believe that the naturalist's view of Nature without any faith spectacle has grandeur to be cherished on its own merits. Neither a Christian nor non-Christian person should be sad, for each gets what he chooses.

Non-Christians do not need Christian spectacles to see and appreciate some of the verities of Christianity, such as the parables of Jesus. These moral writings have been seen and admired by many, Atheists and Christian alike. One person, Thomas Jefferson, a Deist, judged these parables to be special and of value to his overall moral sense and proceeded to tuck them into his life's philosophy. Good morals (Jesus' parables) and good science (Darwin's Theory) are there for all to see in the natural world, and no special spectacles of any kind are needed.

SCIENCE AND GOD

The forward to Falk's book by Francis Collins gives an example of Christians unnecessarily feeling under assault from scientists over evolution:

> *Polarizing influences abound, locking these worldviews in seemingly irreconcilable conflict. Some evolutionary biologists cite growing evidence from the fossil record and DNA analysis to argue that evolution proves there is no God. In the process they commit the logical fallacy of using natural laws to exclude the supernatural.*

Most evolutionary biologists separate the two worlds, the supernatural from the natural, simply because their work in the natural world does not address God in the supernatural world. Falk, however, does not keep them separate and affirms his belief that science, or Nature, is subservient to God:

Scripture describes the activity of God. Science, whether its practitioners realize it or not, also describes the activity of God.

This is Falk's Christian belief. Fine. However, most other scientists do not see God's activity in Nature, for, in fact, they work very hard to be sure that gods, devils, and demons are not involved. In physics, for example, what activities of God are to be seen in the physical processes such as, the radioactive decay by the weak nuclear force of a neutron or a helium 3 atom? What actions can God take with a neutron—twiddle with the weak force, one of the four basic forces of Nature and change the radioactive decay half-life or the decay scheme? Those would be miracles if changed. By starting with the declaration that God has the authority over science, Falk has put himself at odds with the scientific community who think that science (Nature) is independent of the Christian God, the Hindu gods, and all other gods.

Next, Falk claims that academicians (scientists) have a distorted view of creation (Nature) with an argument that Christians can wear "faith spectacles," which allow them to see images that others cannot see.

Most academicians, because they do not own a pair of faith spectacles, see a distorted image of creation.

Possibly we should stop critiquing Falk's book here by declaring that it is a religious book and go no further, for religious books should not be judged by science standards. But since he is addressing college students on science, it is interesting to compare his arguments on evolution with those presented at public universities.

Most scientists are willing to grant that Christians may see things differently through their "faith spectacles" that non-Christians without them do not see. This is, of course, should be considered a gift of religion, to have a different view by virtue of their religion. But Falk errs on calling what academicians (scientists) see without "faith spectacles" distorted when they are wearing no "faith spectacles." The scientist's view of Nature is universal—that is, a scientist sees the same data, whatever religion he may have, or even if he has none at all; Buddhists, Hindus, Atheists, and Christians see the same data, but they may interpret what they see differently.

> *Most academicians, because they do not own a pair of faith spectacles, see an undistorted image of Nature, including creation. Christians have the option to see Nature as an academician without their faith spectacle or as a Christian wearing their faith spectacles and see another image, the Christian image of Nature, including creation.*

Since there are thousands of religions, there would have to be thousands of "faith spectacles," each one giving a different interpretation. Can Christians argue that their "faith spectacles" are better or worse than those of other religions?

GRADUAL CREATION

We need to move beyond this bit of Christian exuberance on "faith spectacles" and anti-theistic arguments and discuss the difficult task that Falk, as a professor, has before him—conveying to students and his church that the concept that Darwin's Theory is an accepted evolutionary theory of the scientific community without appearing to argue against his faith. Falk declares that his task in writing his book about faith and biology is:

To lay before the church the reasons why almost all scientists (including Christian ones) believe in the gradual appearance of life on this earth.

The need for carefully selecting words that do not conflict with church dogma is appreciated, but scientists do not believe in "the gradual appearance of life on this earth." The appearance or creation of our first ancestor was a onetime event, and from there is it has evolved with modifications. We don't know exactly when or where, but the creation of the first replicating bacteria occurred about three billion years ago. Since that time, evolution has produced all living organisms, not by creating new ones, but by modifications of that common ancestor by the processes described by Darwin's Theory of natural selection.

The use of the term "Gradual Creation" for evolution by Falk appears to be an attempt to explain one of the tenets of Darwin's Theory without giving Darwin credit and ignoring the other tenets. However, since Falk's book argues that science is an activity of the Christian God, it falls far short of acknowledging the independence and importance of natural science in describing Nature. The science of evolution is the same for all life forms. It is interpreted differently only because individuals choose to use (in the case of Falk) a distorting "lens of faith." Falk says:

As Christians we view the appearance of (those) cells through our lens of faith. The first cells appeared in response to God's command...

But cells are part of Nature's biological evolution and can only be viewed accurately with natural eyes without any lens. There are many (thousands) faiths (gods), so are we led to believe by Falk that there should be a thousand different appearances of the cell in response to each god? By removing religion from influencing our understanding of cells, Naturalists would simply say:

> *As Naturalists we view the appearance of life as cells with our natural eyes, no lens required. The first cells appeared as a result of Nature's laws of physics, chemistry, and biology through natural selection and has molded them into the living creatures we see today.*

Christians, Buddhists, Hindus, and Naturalists alike, when looking through the same microscope, see the same cell. The Christian difference is only in the religious interpretation of the image. Believing that a supernatural God intervenes in the biological process is only an overlay (an interpretation) on top of what is seen in natural science. This additional layer is a choice made by Christians for Darwinian evolution, which works quite well without the Christian God's involvement. Believing in a God that impacts Nature (physics, chemistry, and biology) is a faith statement superimposed by Christians. Science says, be our guest to your beliefs, but keep them in the supernatural world of religion and out of science laboratories and classes.

All of this may sound scary to Falk's students, for it requires rejecting or changing the biblical Genesis story to the status of a metaphor. It is suggested that teaching Darwin's Theory would be a far easier task for Falk to accomplish than the one where he has to use his Christian spectacles when discussing natural evolution. If it is science, why not use your everyday eyes with no spectacles required? Cannot Christian students take off their "faith spectacles?"

BUILDING A BRIDGE

Many of Falk's students are from Christian high schools, where their biology classes taught the supernatural biblical Creationism theotheory.

In college-level biology classes, these students face exposure to a wider world of ideas, which includes the conflicting specter of godless theories, including Darwin's Theory of natural selection. Addressing this challenge, Falk's approach, as the subtitle of his book states, is by *"bridging the worlds between faith and biology."* Falk defines the bridge he wants to build:

> *Scripture, through the power of the Spirit, is the very bridge that will take us from the world of faith into the world of Nature— the very fingerprint of God. Scripture describes the activity of God. Science, whether its practitioners realize it or not, also describes the activities of God.*

Since scriptures and Nature are, according to Falk, both activities of God, building a bridge from the world of faith (God) to the world of Nature (the fingerprint of God) would be like building from God to God. A bridge needs to connect to two different sides to be credible. But Falk never recognizes Nature as being independent of God, so he never gives his bridge a second anchor point in Nature. Falk's bridge only circles back on itself, for he does not allow it to reach beyond God to the other side of the chasm where Nature resides.

More importantly, the bridge metaphor is simply an incorrect approach to understanding science and religion. They are independent, and building bridges would only mix the two and harm both. Instead of building a bridge that does not connect God and Nature, Falk should build respect for the independence of each, for each has something special to contribute to humankind—but when mixed they both lose their unique identity and the ability to provide different truths and insights.

For example, what could have been learned by building a bridge between God's geo-centric theotheory and Nature's helio-centric (Copernicus') theory? Instead of building a bridge between the two concepts, Copernicus started a revolution by removing religion and declaring that the motion of the planets, the Earth, and the Sun was a science question. The answer was the helio-centric theory that removed the Earth from being the center of the universe and replaced it with the Sun and moved the Earth to be one of the planets orbiting the Sun. Copernicus' natural theory is independent of God

The same is true with the Darwinian Revolution. You either have God as the designer of man, directing evolution, or you have Nature, designing man without a designer. What is needed is to scrap the bridge concept and recognize who has the authority over man's evolution in Nature. Evolution is a natural process, and as such it is under the authority of Nature. This does not preclude Christians believing separately that man's creation is part of a supernatural narrative in which God is the authority.

Although Falk tries to bring peace to the concept of evolution for Christians, he will never get there by only looking through his Christian spectacles. A scientist would paraphrase Falk's title:

> *Coming (as a Christian) to Peace with (godless) Science (by honoring separation from and independence of the natural world of science.)*

God demands belief in his supernatural authority, and Nature demands adherence to the scientific process in the natural world. Separating the two is a good thing for man, and he will be served better by hearing two independent views of man and the universe rather than one muddled one.

AUTHORITY

The unfolding of God's story starting with Genesis is part of a long religious narrative for believers and their God. It is not about science or Nature and does not need verification by the scientific community.

It is a story about the Christian God's message to humanity recognizing its foibles, nobility, and aspirations. As Galileo observed over four hundred years ago, it is religion's job to teach salvation, not science. Had Falk appreciated Galileo's wisdom, he would spend less time trying to build bridges, which don't work, and more time building respect for the independence of religion and science.

Falk does state that the Bible should not be taken as a textbook about science. He quotes John Wesley about scripture: "He (God) made the stars also, which were spoken of only in general, for the scriptures were written not to gratify our curiosity, but to lead us to God." From this quote Falk argues:

> *Maybe our young people do not have to live in two separate worlds*
> *after all. Perhaps modern science and faith can be united within*
> *one world.*

Unfortunately, unifying science and religion (God and Nature) is exactly the wrong thing to do, for it is their independence that matters. Falk hopes that Christians possibly could "see if it might not be God's truth that science has unknowingly been discovering." Again these words give his underlying belief that his Christian God is in charge of Nature and science. Further confirmation of his "God in charge" belief is found in a 2009 statement[77] from his BioLogos organization, which argues that science is just more of God's revealed truth:

> *We affirm that the truths of Scripture and the truths of Nature*
> *both have their origins in God.*

As such Falk is following Dowd's and Collins' approach of thanking God for the truths of both the scriptures and Nature. This combining of Nature and God under God's authority leaves no room for discussion on the independence of science and confirms that Falk is stuck in the old Christian dogma of God having authority over Nature. This failed when the church confronted Copernicus' theory four hundred years ago, and it fails today in applying church dogma to Darwin's Theory. It only continues the "war" with science, in which the only "winner" is Dawkins and his friends, who will sell more books to those who value the independence of science and wish to resist Christians arguing for the dominance of their supernatural God over Nature (evolution).

Christians have great difficulty defending the scriptures as an infallible base of universal knowledge and morals. Time and the accumulation of knowledge have revealed that some of the scriptures are culturally based and have become unacceptable in light of changes in cultures from the time the Bible was written. So for Falk to use the scriptures as his sole reference for the truths of Nature requires that he pick and choose among biblical passages.

Most people today would argue (to use only two examples) that the biblical acceptance of slavery and of killing homosexuals is not moral. Falk unfortunately quotes Phillip Johnson, a lawyer, whose wholehearted support of the Intelligent Design theotheory gives him less than sterling credentials with the scientific community. Johnson was a salesman for the Intelligent Design theotheory, and a major part of his sale pitch was to discredit Darwinian evolution. He has had enough time and money to come up with his best arguments supporting ID, but when put to test in a courtroom in Dover, PA, the ID theotheory he was selling failed scientifically and constitutionally.

Further, Falk again quotes Johnson's critique of Steven Gould and his non-overlapping magisteria argument for science and religion and faults it for providing science with a broad worldview, including the origins and operation of the entire universe. Obviously, Johnson and

Falk think that a broad mandate is acceptable for God (religion) and not for Nature (science). Most scientists are in disagreement with Gould's magisteria arguments, but not with his definition of the scope of science (Nature).

It is argued that Nature does indeed have a broad mandate to explain how man and the universe were created, designed, and evolved in the natural world. But Nature also gives the same breadth of scope to religion, for the two are independent. Why should God and Nature, being separate, not have broad and differing views? The difference is that Nature is the authority in the natural world and religion in the supernatural world. Darwinian evolution explains man's evolution as a process in the natural world. Religion can add supernatural views for believers, but it cannot alter the natural world. Religious views can be meaningful and insightful and provide worthy independent views by God, but they do not describe Nature.

THEOTHEORIES

In Falk's book, Darwin's Theory of natural selection is unacceptable, for it describes Nature's evolutionary process as godless, includes chance (random mutations and natural selection), and operates over billions of years. But Falk knows that Darwinian evolution is a robust theory in science, and cannot be easily discarded. He also knows that earlier Christian theotheories (Creationism and ID), which had been previously offered by Christians to replace or modify Darwin's Theory, have failed. Falk describes why they are inadequate. But instead of turning to Darwin's Theory fully, he feels compelled to veer off and move toward another Christian theo-theory, Theistic Evolution, which has God involved. He accepts one tenet of Darwinian evolution, the long time period, but uses non-Darwinian terms for its description—namely, Gradual Creation. Using this term, he sets out to convince his church and students that man's creation and evolution were not completed in a few days as described in the scriptures, but gradually over billions of years. Supporting this approach, he describes the radioactive dating technique, which is used to determine dates back billions of years, depending on the radioisotope used. Having introduced the natural science estimates for an old earth and an old universe, he has set the stage for discussing life on earth.

However, there is a problem with the term Gradual Creation: it is misleading, for it is really Gradual Evolution he is discussing. Creation of the first living organism, a self-replicating cell, happened only once, and after that, it was Darwinian evolution that gradually modified old species to make new ones. Falk sees the "the parade of life" differently:

> *Life appeared at the command of God, and the parade of life was always under his control. Both the Bible and the biology are entirely consistent with this view of life.*

The first sentence is a belief statement from the Christian narrative, and that is fine. But then Falk mixes his religion with science and gets into trouble. As a scientist he should acknowledge that most evolutionary biologists argue that biblical creation is in total conflict with natural science. Falk must have very strong "faith spectacles" indeed to declare that the Bible and Darwin's godless biology of natural selection "are entirely consistent with this view of life."

Then he goes on to discuss the abundant fossil evidence for Gradual Creation operating on the Darwinian evolution time scale (natural selection acting over many billions of years) and gives God credit for creation, but cannot come to say that it is the evolution of living organisms.

> *What will it take for many Christians to realize that they are missing an opportunity to feel and sense the Presence of God in*

creation? Science studies that creation, since it is God's creation, "surely the Lord is in this place."

More mixing science with the Christian narrative:

The majestic living picture of the interior of the cell is a master-piece like no other. Created by the Word of God's command and the spirit of God's presence, and nurtured by the vision of God the Father for his offspring, the cell is God's painting come to life.

The first replicating cell "created by the Word of God's command" is another string of religious theoscience statements that Falk uses to support his theotheory of God's involvement in natural evolution. Falk makes it clear that God can use miracles when needed.

God can move suddenly in ways that represent unmistakable, but he also works in subtle fashions, in manners that frequently be detected except through the lens of faith.

In his book, Falk gives a strong verdict against the Christian theotheory of Creationism, the biblical description of man's creation, which he calls "sudden creation":

If the most predominant version of "sudden creation" were true, the science of nuclear physics, astronomy, geology and biology would all be utterly wrong. It cannot be taught in a science classroom because it is not science. It is religion.

This critique is exactly what a Naturalist would say, and Falk has taken a step away from Creationism ("sudden creation') and a step toward separating what can be taught in a science classroom, but he cannot see Nature being separate from God.

But Falk finds he has put himself into a conflicted position with many Christians and makes a plea for them not to despair and honor those Christians who remain Creationists:

> *Although you may be absolutely certain that God created gradually, this does not mean that you are somehow less obligated to love and care for someone who is equally certain that God created suddenly.*

This is recognition of an awkward position Falk has made for himself when forced to give obvious conflicting statements about evolution. The way out of this pickle is to recognize that Nature has man and his ancestors evolving gradually while acknowledging that God in his narrative uses sudden creation of man. In this case being open about Nature and God would solve the problem, but since Falk does not seem to be able to speak about Nature's Darwinian evolution[78] he uses God for man's gradual evolution and creates a problem among Christians and scientists.

Falk's fear of offending Christians causes him to exercise excess caution when dealing with Darwin. This he does by never mentioning Darwin's name in his text, nor as a reference. In the description of one tenet of the evolutionary process, the long time scale, he calls Darwinian evolution by an incorrect name of "Gradual Creation" instead of gradual speciation and avoids Darwinian nomenclature such as "natural selection." His biology book on the evolution of man may be the only one written in the last 150 years that does not mention Darwin's Theory or Darwin's name. To the contrary most scientists think Darwin's contributions to biological evolution are worth noting and include:

> *Natural selection, the single tree of life, genealogical classifications, selective extinctions, sexual selection, co-evolution and economy of Nature.*

Darwin was a genius and a giant in biology. Possibly in the future, Falk may be able to use Darwin's name and honor Darwin's sensitivity to his wife's strong Christian beliefs and the conflicts presented by his work in Darwin's household. Darwin was able to appreciate his wife's belief in a supernatural Christian God while independently being an advocate for natural evolution. More importantly, he did not compromise by using Christian miracles to appease his wife's beliefs in order to explain evolution. He let the facts fall as they may. Darwin's science and his wife's beliefs were successfully kept separate in their two different worlds—Darwin in the natural world with real life experimental data, and his wife with her beliefs in the Christian supernatural world. This is a poignant and successful example of separating Nature's authority over science while respecting God's authority over Christianity.

Falk's support of Christian theotheories has been forced to change in light of the defeats suffered by the Intelligent Design theotheory. In a BioLogos article[81] Falk concluded:

> We also need to deal frankly with the notion that the fundamental tenets of the Intelligent Design movement, as laid out so clearly by Phillip Johnson almost two decades ago and developed so articulately by Michael Behe and others have missed the mark. It might have seemed much easier for Christian theology, at least in the evangelical tradition, if they had been right. Our mandate, however, is not to settle on that which is easiest, but rather to pursue what is right. We are committed to doing so within the context of evangelical Christianity.

DARWIN'S THEORY

Although Falk is heading toward the acceptance of Darwin's Theory, he is caught in the awkward position at this time of going only part of the way and receiving fire from both religion and science. His discussion of Darwin's Theory, which he never names, reveals this awkward position.

NATURAL WORLD

Falk's view of the natural world is replaced by the supernatural world, for he uses "faith spectacles" that "see" only the supernatural with its miracles:

> *Commitment to the principle that the supernatural is just as real as the natural. It is the transcendental Presence that set in place the rules scientists study, and it is the same Presence that can suspend these rules anytime, anywhere.*

This is a case of mixing religion with science, for not only does Falk contend that the supernatural world is as real as the natural world, but he goes further: whatever rules (theories) science discovers, God can suspend them—that is, work miracles and change natural science at his supernatural will. This is a case of not wanting to address the natural world and meet Darwin's Theory head on.

LONG TIME FRAME

As a step toward accepting Darwin, Falk takes one of Darwin's tenets, the long time frame for biological natural selection, and incorporates an incorrect term, "Gradual Creation," for evolution, for he feels he cannot use the word "evolution." This definition would have allowed him to bring the concept of evolution over billions of years into the conversation of man's evolution, but it puts him in the indefensible position of trying to explain God using Nature's time frame.

CHANCE

Falk offers "Gradual Creation" guided by God as the mechanism of Nature's natural selection. Ignored within the explanation of Gradual Creation is the critically important role that chance plays in Darwinian natural selection. Chance is the key factor in gene (DNA) variability from gene copying errors and DNA radiation in variable adaptions by natural selection. Thus, chance is critical to making evolution work. Inserting God, who is omnipotent and always certain, into natural selection makes the use of chance or uncertainty of the outcome unacceptable, so Falk leaves chance out and uses Gradual Change instead.

DESIGN

Falk insists that God directs evolution; accordingly, the design of man is by God. With this position he should explain why God is such a poor designer: our jaw is too small for all of our teeth, and the fertilized egg only has a 50 percent chance of attachment to a woman's tubes and producing a fetus. He is also a cruel designer: one of the life forms he designed, the polio virus, is designed to attack children preferentially and the Black Plague bacterium to kill indiscriminately.

But most important is that when discussing science in the natural world, Falk leaves out Darwin's most important discovery: that our common ancestors and man were designed without a designer—that is, by Nature's processes.[80]

MORAL SENSE

Christianity holds not only that God created man, but had him born with an inclination to evil (Original Sin). The Nazarene Evangelical belief statements include:

We believe that man is born with a fallen nature, and is, there-
fore, inclined to evil, and that continually. We believe that the
finally impenitent are hopelessly and eternally lost.

The Christian belief in Adam and Eve and their fall from grace defines that man is inclined to evil. This is opposite of what is observed in experiments on the moral sense of babies and children. Scientists now argue that man in the natural world is born with an evolved moral sense for knowing good intentions from bad ones.

In natural evolution there is not a point where man fell from grace and was suddenly inclined to evil. On the contrary, man's moral sense was shaped by the evolution of caring, altruism, and cooperation of many ancestor species over many millions of years. Such drives by individuals for his family and his tribe helped the survivability of our ancestors. Our basic moral sense is a product of natural evolution and not a gift given by God.

SUMMARY

Falk's book, *Coming to Peace with Science: Bridging the Worlds Between Faith and Biology* is aimed at an audience of Evangelicals who place the authority of the biblical Scriptures over natural science. The book is his attempt to start the exposure of Evangelicals to the wide world of natural science—in this case to Darwinian evolution. This book should have been labeled a Christian book about a Christian theo-theory of evolution, Gradual Creation.

The book is, in effect, a snapshot of the status of the Evangelical viewpoint on natural evolution at a time of transition of their church, from a strict creationist view of evolution to the first halting and limited steps toward the acceptance of natural evolution. The task Falk undertakes in the book is difficult, for he is facing his church's dogma—limiting not only the scope of what he can present, but also the words he can use. For example, the words "Darwin" and "natural selection" were obviously too toxic to use in the book.

Falk's approach to introducing Darwinian evolution is to use the theotheory of Gradual Creation. This comes at a steep price and can be faulted for being misleading as a description of the basic science of evolution. Falk knows this, and this book is the best he could do at the time, for in an article on the BioLogos Web site written well after the book appeared, he opens the door much wider for acceptance of Darwinian evolution by Christians:

> There are two ways in which evangelicals view this story. One is historical, the other, allegorical. To publicly acknowledge that and to make it clear that the latter view does not in any way disengage an evangelical from their faith would be of considerable significance. Let's allow both views to co-exist in evangelicalism for now. I am convinced that we can eliminate the barrier by simply admitting that there are many deeply committed Christians who believe that many elements of the story of Adam and Eve are not historical.

This Web site's article shows, as his book did not, that he has moved down the path of understanding that the views of natural science may be considered, or as he says, co-exist.

It is easy to fault Falk's book for the lack of scientific rigor when inserting religious concepts that short-change Darwin's Theory. It is explainable, for Falk notes that he is using "faith spectacles" when talking about science. If he had labeled his book *A Christian Theotheory on Evolution*, it would fairly reflect his position that God is in charge of evolution as required by his religion and let everybody know this is a religious book.

But the implication that he had crossed a bridge from supernatural religion and found the light on the other side with Nature's godless science is incorrect. Falk's bridge that he crossed was from the Creationism side of his religion to the Gradual Creation side—that is, a bridge with both ends within the Christian religion. That is fine for speaking in the supernatural world. But Nature is in the natural world, still waiting for Falk's comments and connections by a bridge.

The BioLogos Web site gives Falk a different audience with who he can discuss the importance of Darwinian evolution. There he does discuss Darwin's Theory, which he obviously appreciates.[82] Somehow Darwin's deserved credit was omitted.

> *There are few things more certain in the mind of virtually all biologists than the fact that the earth is 4.5 billion years old, that life's diversity has arisen through common ancestry, and that natural selection has played a key role 8 in driving its outcome. Rather, it involves an all-encompassing theory which, following 150 years of testing its many dimensions, is consistent with all the sub-disciplines of biology. One of the hallmarks of a great scientific theory is cohesiveness. Few if any theory in the history of science has ever unified all the disciplines of the natural sciences as has the theory of evolution.*

This is a sound observation of Darwin's Theory. One can only wish that he had started out his book with this as the lead paragraph with Darwin's name included.

To complete his transition to an independent view of Nature, Falk needs only to add that Darwin's Theory operates in Nature, the natural world, without the need for a supernatural God. Separately as an independent supernatural view, Christians can have a God's view of man's creation-evolution.

Scientists are empathic to Falk's efforts to peel away the Evangelical dogma on man's evolution. The critique of his book is but an indicator of how difficult it is to remove the religious dogma that separates the Evangelical version of science from natural science. His "building a bridge" metaphor may have been helpful to move along Christians in understanding science, but it is hoped that Falk will recognize that the real goal is not to build bridges to mix his religion and science together, but to understand that it is their separation and independence where peace can be found. Hopefully his next book on evolution will avoid bridges to nowhere and let his students see a little of the world of Nature without "faith spectacles."

Saving Darwin-Giberson

A fourth Christian reference book, *Saving Darwin: How to Be a Christian and Believe in Evolution,* is by a physicist, Karl Giberson. Certainly Darwin does not have to be saved from/for the scientific community, for his theory is settled science and a cornerstone of modern biology. *Saving Darwin* for Christians is Giberson's view on the conflicts between his Christian views and those of natural science. As with the previous three authors, Giberson states that his Christian God is the authority over evolution. From that position, he leads his readers down a well-trodden Christian road, which, unfortunately, leads to his proposing to insert God into Darwin's Theory and accept a God-centric theotheory, which has already failed.

Giberson is a physics professor and a science historian, and his book takes us on his journey from his discovery of science to his present dilemma of reconciling his Christian faith with Nature's Darwinian evolution. Giberson gives an in-depth and thoughtful analysis of his transformation from a Fundamentalist Christian to an Evangelical Christian. He then gives a brief history of Darwin's life and his transition from a Christian in his youth to an Agnostic later in life.

After Darwin's time, the Victorian era, which once embraced a strong Christian society, changed into a society that would even allow the discussion of the "burial of God" by the end of the century. Christians argued that philosophers such as Hume and his ilk were as dangerous as Darwin in causing the perceived (by Christians) downward thrust of society's morals at the end of the century. Science was beginning to accept the potential fundamental understanding that Darwin's Theory of natural selection would bring to all of biology.

Giberson discusses the birth of Fundamentalism, the Scopes trial, and the subsequent early series of theotheories offered by Christians in the hope of having their God seen as the authority over man's evolution. He leaves Christians feeling that Darwinism is not going away and that Christians need to find newer arguments to keep God as the authority over man's evolution. Giberson's Christian belief dims his ability to observe that the scientific revolutions have removed God's authority over Nature and that God and Nature inn the future must be considered to be separate and independent.

Giberson's book takes an unnecessary sidetrack into social Darwinism, which is interesting, but not at all important to the central issue of God's involvement in man's evolution. We will act as Darwin did in his time—stay focused and uninvolved in side interpretations of evolution. Along the way we will assure Giberson that his fears of Darwinism becoming a religion are unfounded.

Giberson gives an excellent, detailed discussion of the unfolding series of theotheories—Creationism, Scientific Creationism, and Intelligent Design, which were offered by Christians to modify Darwin's Theory in order to keep God involved in evolution, particularly in science courses in the public schools. However, these Christian theotheories fell by the wayside as they were rejected by the scientific community and defeated in court cases. Each failed court case represents an attempt to inject a Christian theotheory into the public school science classes. Giberson's descriptions of the court cases are interestingly told.

Without having much success in having their theotheories accepted by public schools, Christians became frustrated and fearful of the continuing advances made by science. Giberson has the Christian fears summed up by D. James Kennedy, a Fundamentalist, who is well known to blame Darwin for everything bad in the world:

If evolution is true then we are simply the product of time and chance, and there is no morality and no intrinsic worth to human life.

This gloomy prediction has, of course, been proven wrong in Christian countries and countries of many other religious persuasions. Most scientists, including Giberson, declare that:

there is simply too much evidence in its {Darwin's} favor.

Understanding the scientific importance of Darwin's Theory and its impact on science, Giberson explains that Darwin's Theory created two revolutions impacting Christian views: first, he notes that the biblical creation theotheory in Genesis (man created by God in final form) is rejected:

The destruction of the traditional {Christian} creationist picture, where God created all things via individual supernatural acts more or less as we find them today.

The second of the two revolutions establishes Darwin's Theory as a natural process by which man and all of life evolves:

The establishment of random, purposeless selection process, natural and sexual, as the only creative mechanisms at work in natural history.

Giberson continues with his overall view of Darwin's Theory but quickly injects God in a similar fashion as other Liberal Christians (Dowd, Collins, and Falk):

I think evolution is true. The process is an expression of God's creativity, although in a way not captured by the scientific view of the world.

After declaring that evolution is true, Giberson continues to see only with his Christian lenses. This puts him in a pickle, for he has presented himself with a dilemma—Darwin's Theory is true using Nature's processes, but he declares that the processes are really God's supernatural process. In effect, Giberson is attempting scientific robbery in the name of God—steal Darwin's Theory and replace Nature's process (natural selection) with God's actions (take no chance and design all life) and call it God's creativity. Nature's natural selection has worked for billions of years before Giberson attempted to move Nature out of the process and place God in as the creator and designer of man, beast, and pond scum.

By this act Giberson misses Darwin's major idea in evolution, that living organisms are designed by Nature's process—designed without a designer. By omitting this key tenet of Darwin's idea, Giberson reverts to the old and discredited concept of God being the intelligent designer, an old, failed theotheory that violates some of Darwin's major tenets.

Recently the long arm of evolutionary explanation has reached directly into territory where traditionally religious phenomena reside. Evolutionary theory now provides naturalistic explanations for altruism, morality, our religiosity and predisposition to believe in God, even the love we feel for our children.

His strong desire to defend God's role in evolution can be seen in his negative reaction to a publication, *Statement on Teaching Evolution* by the National Association of Biology Teachers (NABT), which he thought diminished religion and favored science. It was particularly troubling to Giberson that the NABT's article described Darwinian evolution as being unpredictable, unsupervised, and impersonal. Immediately, cries were heard from Christians seeking to change these words and keep God involved. Although the words "unsupervised" and "impersonal" were subsequently removed from the teachers' report, the scientific content and intent remain unchanged. Giberson's problem is his continuing to cling to his supernatural God's involvement in evolution. The words "unpredictable," "unsupervised," and "impersonal" are all true, if somewhat folksy. Nature is the causal agent for natural selection and employs chance (unpredictable), designs with a without a designer (unsupervised) and does not employ a God (impersonal). The NABT report was just another report that heightened his fears of science diminishing religion.

Giberson is in for many shocks coming from the deeper understandings of the natural evolution of morality. He notes his fears from E. O. Wilson's comments on scientific materialism replacing supernatural religions. Wilson notes:[83]

> *If religion, including the dogmatic secular theologies, can be systematically analyzed and explained as a product of the brain's evolution, its power as an external source of morality will be gone forever and the solution of the second dilemma will have become a practical necessity. What I am suggesting, in the end, is that the evolutionary epic is probably the best myth we will ever have. It can be adjusted until it comes as close to the truth as the human mind is constructed to judge the truth. And if that is the case, the*

mythopoetic requirements of the human mind must somehow be met
by scientific materialism so as to reinvest our superb energies.

Giberson could be informed by Patricia Churchland's discussions of what neuroscience is telling us about the evolution of morality. Wilson's projections that religion may be proven to be a product of the brain are being borne out by the neuroscience research today. As previously noted, I also believe that Wilson underestimates the stubborn tenacity of the human mind to embrace supernatural religious narratives. Scientific materialism may be a logical outcome of Wilson's arguments, but his logic will not be a totally satisfactory replacement for some human psychological needs for the supernatural anytime soon, if at all. A percentage of believers want to embrace the supernatural regardless of validity.

Giberson's book is an expression of his deep concern for keeping God as the authority over creation-evolution:

Can there be any role at all for God in our own creation story?

Giberson is arguing either/or and has not recognized that he can have both if he can separate God's world from Nature's world. With their independence and separation, he could then find that God's creation story and Nature's creation story are both available separately; one in the supernatural world and one in the natural world. His problem has been self-created and is self-defeating, for by mixing religion and science together, he rejects science in the natural world.

CHRISTIAN THEOTHEORIES

Over a thirty-year period, Giberson has observed the Christian attacks on Darwin's Theory and understands the arguments from both the scientific as well as the Christian view. He has witnessed the rise and fall of a number of Christian theotheories and his book provides an excellent historical narrative of the proponents and opponents of various theotheories (Creationism, Scientific Creationism, and Intelligent Design) and discusses why they have been unsuccessful in competing with Darwin's Theory. A few of the highlights from his insightful historical narratives of theotheories are outlined below.

SCIENTIFIC CREATIONISM

In the early part of the twentieth century, William Jennings Bryan conducted a national campaign to rid the country of Darwin's Theory, which was beginning to be taught in public schools. A 1923 book, George McCready Price's *The New Geology*, was seized upon by Christians eager for the support of scientific sounding arguments for Creationism. Price dismissed any geology that did not fit his view of what the scriptures said. For example, he rejected the geological premise of finding older layers the further one goes down in rock formations, giving some order to fossil dating, he dismissed radioactive time-dating of rocks, and he argued that dinosaurs and man lived at the same time. The book contained many mistakes, but instead of being corrected, Price was hailed as "the expert" in geology by Fundamentalist Christians looking for arguments against Darwin's evolution. Bryan loved it. But it ran out of steam as scientists and others brought the errors in the book to light.

A second book the Fundamentalists grabbed was *The Genesis Flood: The Biblical Record and Its Scientific Implications* by Morris and Whitcomb in 1961. The authors were aware of many of the errors in Price's book and set out to correct them, but in the process, they added extensive biblical interpretations with additional science errors. The book announced that

> *We desire to ascertain exactly what the Scriptures say. We do this*
> *from the perspective of full belief in the complete divine inspiration*

and perspicuity of Scripture, believing that a true exegesis thereof
yields determinative Truth in all matters with which it deals.

In 1982, the *McLean v. Arkansas Board of Education* lawsuit ruled against teaching the Scientific Creationism theotheory in public schools. It was ruled to be a religion and not natural science. The efforts to get the Creationism and Scientific Creationism theotheories into public schools have not fared well. Subsequently, Scientific Creationism lost again in the 1987 court case *Edwards v. Aguillard*, when the US Supreme Court ruled that a Louisiana law requiring that the Creation Science theotheory be taught along with Darwin's Theory in public schools violated the constitutional separation of church and state. The approach of mixing religion and science failed for being bad science in the scientific community and in the courts for violating the separation of church and state provisions of the Constitution.

By the 1990s, as no scientific experiments had ever been forthcoming in supporting Scientific Creationism, it had become a scientific joke, consistently providing raw material for television comedies for posing as science with no scientific basis.

INTELLIGENT DESIGN

After Creationism failed to make inroads in the public schools, other Christians picked up the challenge and proposed another theotheory, Intelligent Design, which proposes that at times Darwinian evolution works (intra-species), but does not work at other times (inter-species), for to create new species it is necessary to have an Intelligent Designer, God. ID received significant support from private money[84] and had significantly greater publicity than Creationism leading Christians to expect success in their court battles.

Scientists argued that ID real support was scientific ignorance and they came out in force to oppose ID introduction into the public schools. Even after much support by Phillip Johnson and the Discovery Institute, ID received a bashing in a Dover, Pennsylvania, courtroom[85] when the judge's verdict called ID nothing but Creationism in other clothing and ruled against the school board trying to teach ID in public schools.

It was a painful defeat for Christian advocates and brought forth an admission by Giberson that "although I wish it were true, it must be rejected." For Giberson, it was a complete failure:

> *Creationism and ID have made little progress, despite the decades of huffing and puffing and blowing on the house of evolution.*

> *From my perspective ID must be rejected on two completely separate grounds. In the first place ID doesn't work scientifically.*

Finally, the technical support for the ID theotheory proved essentially nonexistent after more scientists looked into its claims that man's biological complexity could have been designed only by God. A specific example of science refuting the ID complexity argument is Russell Doolittle's research (see Appendix D) on the evolution of the complex natural process of blood clotting, which explains the evolution of the process over the last five hundred million years.

The history of the rejection of ID gives an insight into the pain Christians are willing to endure in attempting to give God the authority over Nature in evolution. After failing the science test and as the political character of ID became apparent, ID has become a dying theotheory, but as they say, old theotheories never die, they just fade away. Unfortunately, this one has not totally faded away.

THEISTIC EVOLUTION

Giberson faults the ID theotheories for having God intermittently inserting supernatural acts (miracles) into the process. He then explains that he likes another theotheory, Theistic Evolution, because it uses God in an uninterrupted evolutionary process:

> It {Theistic Evolution} does not require that God interrupt the natural course of events to "occasionally" impact evolution, rather it views God as inhabiting the entire process.

Here again Giberson is using his Christian lenses when viewing a scientific process. Having God "inhabiting the entire process," or any part of the process, is a violation of Darwin's evolutionary process—random variations being acted on by natural selection with no involvement by a God. If God were involved, he would have to be the designer, and no chance or uncertainty would be involved.

The Theistic Evolution theotheory being offered by Christians is facing the same failure mode as previous theotheories, Creationism and Intelligent Design, for it also mixes religion and science, thereby failing the constitutional test of separating religion from the state and failing the natural science test posed by the scientific community.

DARWIN'S THEORY

In the beginning of his book, it appeared that Giberson was on the verge of fully accepting Darwin's Theory ("I think it is true"), but he veered away. from a full-hearted acceptance by insisting on inserting God into evolution as the designer. He has not been able to accept that Darwinian evolution works without his Christian God, for he sees evolution only through his Christian lens:

> As a believer in God, I am convinced in advance that the natural world is not an accident and that, in some mysterious way, our existence is an "expected" result.

Clearly Giberson's beliefs, as much as he admires parts of Darwin's Theory, are such that he cannot accept evolution without his God:

> *Christians have found ways to make it {Darwin's Theory} a part of their understanding of God's creative process.*

By modifying Darwinian evolution to include *God's creative process* and removing chance, Giberson, in effect, is supporting the Theistic Evolution theotheory described by Collins, the Vatican, and others, and not Darwin's Theory.

NATURAL WORLD

Giberson sees the mechanics of Darwin's Theory, which he and most other scientists approve. But since it conflicts with the scriptures, he puts on his Christian lenses and sees the need to inject his supernatural God into Darwin's Theory.

LONG TIME FRAME

Giberson has no problem with the billions of years as the time frame of evolution.

CHANCE

Giberson cannot bring himself to accept that chance is an integral part of the evolutionary process.

> *I am not convinced that we are "accidents results of random purposeless natural selection processes.*

Giberson sees God using Darwin's Theory, so he refuses to believe that we are products of chance. Unfortunately for Giberson, man is an "accident," for chance is a part of Nature's random natural selection process. By removing chance, his statement on natural selection is a religious one, not a scientific one.

Design

Giberson continues to believe that God is the director and designer of man during evolution:

> *Christians have found ways to make it {Darwin's Theory} a part of their understanding of God's creative process.*

Saying that God's creative process includes Darwinian evolution, Giberson is giving the authority of evolution to God and not Nature. This is unfortunate, for many Christians, including Giberson, have not accepted the science of Darwin's greatest idea—that man and every living creature before him were designed without a designer.

Common Ancestor

Giberson accepts this.

Moral Sense

Giberson sees Darwin's Theory as presented by scientists as a cruel process, dooming man's morals:

> *{With Darwin's Theory} a product of a random, purposeless, cruel process, and atheism, moral anarchy won't be far behind."*

This, of course, is a gloom-and-doom prediction without a base of supporting data. It has been over 150 years that Darwin's Theory has been around, and moral anarchy has not overtaken the countries in the Western world.

Calling natural selection "a cruel process" is misunderstanding Nature's process of biological selection by attaching a value to natural selection.

It is the process that got us here, and not many would agree that today moral anarchy has overtaken the countries in the Western world. Finally, there is no link between moral behavior and one's religion, for one sees Atheists and believers of other religions equally as moral as Christians.

CONCERNS

Giberson says there a number of questions that concern him. First, there are questions that are important to him that God, not Nature, can answer. That is, of course, a good reason for him to be a Christian and search for answers from the supernatural world of his God. Natural science does not claim to provide answers to questions outside of science, and certainly not those in the supernatural world of religion. God's words have been and are currently available from the scriptures to provide Christian moral answers for Giberson and other Christians whenever they choose.

It may be that science cannot be helpful looking for or answering the "higher" questions posed some people. But searching for a "higher" answer is something scientists do even when not knowing whether a "higher" answer is possible. Not knowing where it all leads or whether there is a purpose is acceptable uncertainty for scientists. But for those who want more certainty, direction, and purpose, as Giberson may want, he has his religion. Being both a scientist and a Christian gives him two worlds for answers—Nature's world with science and God's supernatural world with religion. Each separately can give him informative answers, but mixing the two together only produces unsatisfactory conflicts.

GOD'S AUTHORITY

Christians have their own supernatural narrative covering the creation of man, but science has natural Darwinian evolution. Science says "don't step on our turf, for you have no authority in scientific evolution." Such comments by science are painful to Christians, for it implies a limit to God's power. However, science's independence is necessary to define the boundaries of the disciplines of science, religion, and democracy. Simply, religion has over-reached its authority, and natural evolution is one of those areas where Nature is the authority in the natural world. Christians are not asked to change their supernatural narrative with its creation story or any other subject, but they are asked to keep their supernatural religious version separate from science's natural version. Giberson fear that removing God from man's moral evolution, as Darwin's Theory and neuroscience research do, will open the doors to Atheism or "moral anarchy." This is a fear common among many Christians, but it has no basis, and in fact, the opposite is probably the case.

Accepting the independence of science, and accepting Darwin in the natural world, will remove the need by Christians to invent further supernatural theotheories to explain conflicts with natural science. This will avoid future theotheory failures and continued rejection by the scientific community that have befallen Christian to date. Scientists offer, possibly too forcefully at times, example after example of conflicts with the misuse of science in Christian theotheories. They should be saying, "devote your time to humanitarian services, which you do well. Surely fighting and losing over natural science theories over and over indicate that you are clearly outside of your field of competence."

In short, supernatural theotheories are not viable alternatives to natural science theories for describing how Nature does her work, including evolution. Religions are a viable force for supplying humanitarian services and giving a critical view of science's activities from a religious viewpoint.

It is hoped that recognition of science's independence from religion will remove any fear that science will ever be a religion, for it recognizes that religion is also an independent discipline and, as such, will remain independent and continue to service mankind as it does today.

SCIENCE AS A RELIGION

Giberson expresses concern with the possibility of science becoming a religion and replacing Christianity. He worries:

> *In addition to lurid expositions of a wide range of scientific concepts from DNA to consciousness, we find suggestions that science should replace religion.*

Giberson's use of the word "lurid" expresses the concern, possibly fear, of science becoming a religion. It is most troubling to Giberson that many of the scientists who are winning the arguments on evolution using natural theories against Christian theotheories are not only good scientists, but good writers and communicators, such as Richard Dawkins and E. O. Wilson. They both explain Darwinian natural selection as a random, purposeless, and valueless process—a fearful vision that leads Giberson to believe what Wilson proposes:

> *ultimately intends that religion will be replaced by science.*

And Giberson notes that if that happens, the door to Atheism will be opened, and the people will feel that they are:

> *a product of a random, purposeless, cruel process, and atheism, moral anarchy won't be far behind.*

Most others do not see Wilson's writings advocating replacing religion with science—far from it. Science is not and can never be a religion. Nor should it be, for science has nothing to say about an individual's values and purpose in life. How appealing and inspirational would it be for science to be a religion with the following selling points?

Science doesn't know how matter prevailed over anti-matter.

Science estimates that the Earth will disappear in four to five billion years.

Science doesn't know how the future of the universe will work out.

Science knows little about most of the material (Dark Matter) or energy (Dark Energy) of the universe.

How many people would be attracted to a religion that has a lot of "how little it knows" and not much of "I know the way" as the guiding light? I would suspect very few.

Giberson's fears that science is conducting a "war" that will ultimately replace Christianity as a religion are unfounded. Science is only trying to peel off old Christian dogma so it can do its thing independently. Religion is only being asked to give up what it should not have had authority over in the first place: the authority over scientific theories. Religion has its role to play independent of science, but that role does not include dictating how Nature's processes work in the natural world for it is Nature, not God, that describes the natural world.

Religion has its role, described by God, in the supernatural world. The Christian narrative with its creation story is not being rejected, but is being removed from Nature and placed back into the supernatural world of God from whence it came.

In his book, Giberson states his frustration for the lack of God's authority being recognized in evolution. At one time his feelings boiled over unexpectedly with an attack on the scientific community:

> They (scientists) *admit they will defend and even promote preposterous notions, rather than admit that God might have some relevance to understanding the natural world.*

To prove his point, Giberson quotes a leading geneticist, Richard Lewontin, who he says has stated this commitment with candor: science (materialism), Lewontin says:

> *cannot allow a Divine Foot in the door {of science}.*

Giberson misreads what Lewontin is saying—namely, that a supernatural God has no right to step into the natural world, so keep it out. This includes the Christian God and any of the other thousands of Gods.

Gods in the supernatural world should not be held to the scientific standards science applies in the natural world. Gods should be free in their supernatural world to provide inspiriting narratives to believers.

SCIENCE'S ANSWERS

In evolution, Giberson sees the beauty and sweep of science. But the same beauty and sweep is also seen by non-theists, including Atheists. The reason that beauty is seen in science is that it is universal and open to all and works for all. Giberson teaches about Darwin and science theories in relationship to Christian beliefs and notes:

> *Science provides a partial set of insights that, though powerful, don't answer all of the questions.*

THANK EVOLUTION FOR GOD

This is a point where Christians should look to their religion for some of the answers. Although the scientific insights into evolution have proven to be powerful as those of particle physics and other branches of science, there are no claims by scientists that science has all of the answers.. In fact, scientists are saying that in many cases, they don't even know if they are asking the right questions.

Scientists are happy with Giberson's recognition that Darwinian evolution is a "natural course of events," but when Giberson argues that God "inhabits" (that is, directs) evolution, there is disagreement. Inserting God into Darwinian evolution in order to remove chance and uncertainty makes it a theotheory, a religious belief, and no longer science. As a Christian, he desires to have his God as the authority over Nature and to have the Theistic Evolution theotheory replace Darwin's Theory. The title of his book should have read:

> *Saving Darwin: How to Be a Christian and Believe in Evolution {if Darwin's Theory is modified with God replacing Nature's godless process of natural selection}*

SUMMARY

As a Christian, Giberson refuses to accept that Darwin's Theory works without the involvement by his God.

> *As a believer in God, I am convinced in advance that the world is not an accident and that, in some mysterious way, our existence is an "expected" result. No data would dispel it. Thus, I do not look at natural history as a source of data to determine whether or not the world has purpose. Rather, my approach is to anticipate that the facts of natural history will be compatible with the purpose and meaning I have encountered elsewhere. And my understanding of science does nothing to dissuade me from this conviction.*

When wearing his Christian lenses, he is comfortable in having God's supernatural world controlling Nature—"our existence is an expected result and no data would dispel it" are simply statements of belief. This statement expresses the depth of his belief and his inability to separate God's supernatural world from Nature's world. He appears unable to view the world without his Christian lenses, which makes his book a collection of uncomfortable struggles as he attempts to explain godless science's successes in many fields. As a physicist he knows that this is not just about Darwin. Einstein's relativity theory allows us to calculate the deflection of light by mass without evoking a god. The energy-mass theory allows us to calculate the energy released from a certain mass of fissionable matter (the atomic bomb works by this theory) without a god. Further, he knows that in the wacky quantum world, uncertainty is king, a world that no self-respecting god would inhibit.

Darwin's Theory is his main concern, for if man's evolution does not need his Christian God, then Christians must pass the authority of man's creation in the natural world from God to Nature. This is a difficult transition to see when you can only see with Christian lenses. The natural world is a place where no lens would work the best.

The recognition of Nature's authority over the natural world is neither harmful nor fatal to religion. Just the opposite—religion will be liberated from the false theotheories of God's control over Nature and allow Christians to focus their energies on tasks within their authority, such as improving the human condition, an important thrust of Christianity, as the Bible notes. Arguing with the scientific community on Nature's turf and losing battle after battle should be a clue to change one's tack—take off the Christian lenses occasionally, dump the theotheories, and enjoy the two views.

There is much the Christian community can do to clear up the conflicts—start by keeping the public schools free from religious intrusion into science, work with the scientific community on the extent of each's authority, and stop using theotheories (Creationism, ID, Sacred Intelligent Design, and Theistic Evolution) outside of the religious community. Finally, have Christians support the teaching of Darwin's Theory in public schools as the keystone of biology. Running Darwin out of the country is an old Christian theme from the 1920s which has failed. It is time for Christians to recognize his importance to the natural world.

Giberson's well written discussion of the history of evolution gives due recognition to Darwin's scientific accomplishments, Darwin's name, and the fact that Darwinian evolution is the cornerstone of all biology. That's progress for Christians to see. Hopefully he can take the next step and accept Nature's authority over science in the natural world and recognize that Nature does not subtract from God's authority in the supernatural world. With this step he will be able to enjoy seeing the two different views of the world, one from his God and separately one from everybody's Nature.

What's So Great About Christianity?-D'Souza

inesh D'Souza is a Christian writer. His book *What's So Great About Christianity?* is a defense of Christianity from purportedly anti-religious arguments and actions by Atheists, scientists, democrats, and public educators who have and are now causing the diminishment of Christianity's greatness. Front and center leading his invented attackers are the Atheists, whom he blames for conducting a "war" against Christendom in which they "cowardly" use the advances of natural science, philosophy, secular governance, and public education to inflict their views on Christians. He is bothered by the fact that in the public schools, natural science with its godless theories is taught; from the biological theory of natural selection by Darwin to the cosmological theories on the Big Bang creation of the universe by physicists, D'Souza feels he must counter these non-theistic subjects by attacking them. His attacks are those of a militant Christian aggressively pursuing his religious beliefs at the expense of natural science.

D'Souza argues that the Christian God is in charge of Nature, therefore, all of natural science is an expression of his supernatural

God. His approach is to mix religion and science and attack all that do not conform to his specific Christian worldview. Thrust into the middle of his attack on religion is Darwin's Theory of natural selection, a secular theory supported by many: scientists, believers of many religions, and Atheists. D'Souza chooses to focus Atheists in this "war."

> *It is a war over religion, and that it has been declared by leading Western Atheists.*

ASSERTIONS

D'Souza defines his position on the greatness of Christianity with seven assertions. Three of these (numbers 2, 3, and 4) address natural science and Darwinian evolution with false assertions. These three key assertions are listed below in italics with each followed by a summary reply (normal script):

2. The latest discoveries of modern science support the Christian claim that there is a divine being that created the universe.

> <u>False assertion</u> - Although the details of the cause and mechanics of the creation of the universe are not known at this moment, modern science tells us that the creation of the universe was most probably by Nature, and an intervention of a divine being is not required. Christian claims for creation are clearly supernatural stories, which can be used by believers as they see fit for sermons, but they do not satisfy the tests required by natural science to establish a known cause. Natural science simply does not support divine claims of any kind, for there is no data to show otherwise.

3. Darwin's theory of evolution, far from undermining the evidence for supernatural design, actually strengthens it.

> <u>False assertion</u> - Darwin's Theory tells us that the design of all biological life, including man and his ancestors, over billions of years of evolution, has been by Nature, and Nature processes do not use or need the intervention of a divine being for Nature designs without a designer. In reality, gods should not be offered as the designer of life forms, particularly man. It would only paints him as a terrible designer, for man is a collection of bad designs that no self-respecting God would want to claim.

4. There is nothing in science that makes miracles impossible.

> <u>Confusing assertion</u> - There is nothing in natural science that makes miracles impossible or possible, for science simply does not address or need miracles. Supernatural miracles cannot exist within natural science. Miracles have a home only in supernatural religious narratives and other stories written by man, where they are often used as literary devices advertising the power of a leader or a god.

ATHEISTS

Who are the Atheists D'Souza readily attacks? Literally they are nontheists, ones without a belief in his supernatural god. There is no Atheist's Holy Scripture that can be followed, so the belief of an Atheist is an individual belief constructed from arguments on Nature. For most of the last two centuries, non-theists, which include Atheists, have been only a few percent of the population. Today the number of non-theists as a group is approaching 20 percent of the population even in our religious country.

D'Souza starts his arguments by labeling his perceived key Atheist themes in their "war" on Christianity. The first theme is that the:

. . . distinguishing element of modern atheism is its intellectual militancy and moral self-confidence.

A second major theme of Atheist discourse is the historical crimes of religion. The Crusades, the Inquisition, the religious wars, and the witch trials all feature prominently in this moral indictment.

As a minority, Atheists must work with the Christian majority in our democracy to support their interest, such as the separation of church and state. An example is secular public education, which is supported by Atheists, non-theists, and many of the Christian majority, but opposed by a minority of Christians who actively seek to inject their God into secular public education.

D'Souza claims that the perils of atheism on a worldwide scale can be seen in the decrease of the numbers of Christians in Europe and believers in America:

Regular church-goer numbers in Europe, depending on the country, is between 10 and 25 percent of the population. Only one in five Europeans says that religion is important in life. Czech president Vaclav Havel has rightly described Europe as "the first Atheistic civilization in the history of mankind." Still, some 40 percent of Americans say they attend church on Sundays. More than 90 percent of Americans believe in God, and 60 percent say their faith is important to them.

There are many reasons for the large increase in secularism and Atheism in Europe and for the lesser increase in America. The main reason has been the recognition of the contribution to humanity from the advances in secular science, philosophy, and governance which have been achieved without the involvement of a God, not by waging a "war" on God. However, D'Souza chooses to ignore secular advances and instead invents scenarios that accuse Atheists of a wide range of sinful activities against religion:

> *For Atheists, the solution is to weaken the power of religion worldwide and to drive religion from the public sphere so that it can no longer influence public policy.*

> *But how should religion be eliminated? Our Atheist educators have a short answer: through the power of science.*

> *Atheists do not bother to disbelieve in Baal or Zeus and invoke them only to make all religion sound silly. The Atheists' real target is the God of monotheism, usually the Christian God.*

These are accusations unsupported by facts. D'Souza forgets that in a democracy religion is separate from government and an independent activity in the public sphere. Religion has no place in public school science classrooms for they are funded by the public and, accordingly, must remain free of religious involvement.

He continues to expand his lists of the supposed evils of Atheists:

> *The Atheists no longer want to be tolerated. They want to monopolize the public square and to expel Christians from it. They want political questions like abortion to be divorced from religious and moral claims. They want to control school curricula so they can promote a secular ideology and undermine Christianity. They want to discredit the factual claims of religion, and they want to convince the rest of society that Christianity is not only mistaken but also evil. They blame religion for the crimes of history and for the ongoing conflicts in the world today. In short, they want to make religion—and especially the Christian religion—disappear from the face of the earth.*

It is a false accusation that public education undermines Christianity. Christianity is one of many religions that are separate from publicly funded schools and it, like all religions, must rise or fall on its own merits and contributions to society outside of government support.

D'Souza accuses teachers, professors, and public education with harboring Atheism and indoctrinating schoolchildren with secular and naturalistic philosophies with anti-religious agendas:

> *It seems that the Atheists are not content with committing cultural suicide—they want to take your children with them. The Atheist strategy can be described in this way: let the religious people breed them, and we will educate them to despise their parents' beliefs. So the secularization of the minds of our young people is not, as many think, the inevitable consequence of learning and maturing. Rather, it is to a large degree orchestrated by teachers and professors to promote anti-religious agendas.*

He further extends the sins of educators:

> *Children spend the majority of their waking hours in school. Parents invest a good portion of their life savings in college education to entrust their offspring to people who are supposed to educate them. Isn't it wonderful that educators have figured out a way to make parents the instruments of their own undoing? Isn't it brilliant that they have persuaded Christian moms and dads to finance the destruction of their own beliefs and values? Who said Atheists weren't clever?*

> *Today, however, we read books like Susan Jacoby's* Freethinkers *that celebrate the fact that we live in a mostly secular society. We find Sam Harris insisting that it is quite possible to develop*

morality independent of the Christian religion or religion in general. We read Theodore Schick Jr. in Free Inquiry *insisting that philosophers as different as John Stuart Mill and John Rawls have demonstrated that it is possible to have a universal morality without God.*

D'Souza again cannot accept the separation of the church from government and secular public education. From this broad range of evil activities by Atheists, D'Souza concludes with an assumption that has no support:

The Atheists' real target is the God of monotheism, usually the Christian God.

D'Souza treats Atheists as an organized threat that acts as one—which, of course, has never been the case. Atheist are individuals who have their own thoughts on a problem and as citizens vote in our democracy.

Atheist is a poorly defined and understood word, so it is easy for Christian bigots over time and D'Souza now to accuse them of many things. The 9/11 religiously inspired attack in New York was initially blamed on Atheists by leading Christians such as Jerry Falwell and Pat Robinson. Further, to blame Atheists for the mass killings of Stalin or Hitler in the past is a false accusation. Stalin and Hitler believed in the value of personal power manifested by their actions to increase their personal power by whatever means. Theism or Atheism was no consideration for they were only traits to be manipulated in order to increase personal power. Instead of a belief in a personal god, they believed in personal power.

NATURAL SCIENCE

D'Souza declares that natural science is a product of Christendom, but he fails to recognize that science and the advances of science have occurred in many countries other than Christendom or Christian Europe:

> *Yet science as an organized, sustained enterprise arose only once in human history. And where did it arise? In Europe, in the civilization then called Christendom.*

History has shown that science is a universal activity of all peoples and its knowledge has been advanced by individuals in countries throughout the world before there was a Christian church. There were Greek and others before Christ and afterwards. Muslim scientists and mathematicians in the Middle East made many scientific contributions, as well as scientists of various Eastern religions in India and China. The early phase of the Scientific Revolution in the fifteenth century included the restoration of the natural knowledge from the Greeks, which had been largely lost, but which reappeared later with the transference of scientific data and advances from Muslim scientists into Europe. The excellent planetary, novae, and stellar observational data recorded by Muslims between eighth and thirteenth centuries made its way to Europe and were subsequently referenced by Copernicus in the formulation of his heliocentric theory of the cosmos. Science prospered in the West from the rise of middle-class and educational institutions, many supported by the church, which allowed more opportunity for hopeful scientists to receive an education and be heard.

D'Souza incorrectly puts the Catholic Church as the driving force for the scientific advances when in reality it was the individual Catholic and Protestant contributors who made the advances, and in many cases, their advances were in spite of the church bureaucracy (Copernicus, Galileo, etc.). The largest institutional church, the Vatican was, at times a damper on any advance of science, for as an organization its highest priority was to defend its dogmas, a source of its power, against any new theories. The advancement of natural science and everything else that was new was a distant second. Science advanced in spite of Christian dogma, not because of it:

> Science was founded between the thirteenth and fourteenth centuries through a dispute between two kinds of religious dogma. The first kind held that scholastic debate, operating according to the strict principles of deductive reason, was the best way to discover God's hand in the universe. The other held that inductive experience, including the use of experiments to "interrogate nature," was the preferred approach. Science benefited from both methods, using experiments to test propositions and then rigorous criticism and argumentation to establish their significance.

Instead of religious dogmas, it was the observational data from the early astronomers, including Muslims, and Christians, that launched science in the West. Later, Galileo's observations with his newly built telescope gave scientists additional data with which they could test new revolutionary theories about the planets and the cosmos. One was Copernicus' helio-centric theory which conflicted with the church's geo-centric theory. The Vatican resisted accepting Galileo's astronomical data, his books, and his support for the helio-centric theory. But the church was not able to stop other astronomers from gathering data and proposing new theories, for the science genie was out of the bottle for all to see, to use and to serve as a base for inventing even more new theories.

Most scientists were Christian, and many saw themselves as achieving God's purpose by going beyond God's biblical words and exploring his creation. They were careful to collect data as observed without injecting God into it, and it was the expanding availability of experimental data that was used to support Copernicus' helio-centric theory and new theories by Kepler and Newton. The new theories did not use a divinity to explain the motions of planets in the heavens.

In America, experiments in the clouds and on the Earth by the less than religious Benjamin Franklin showed that lightning, once viewed as a sign in the heavens from God, was a natural electrical discharge that could be readily demonstrated on Earth. Natural science was becoming secular simply because the data and the theories did not require a God.

COSMOLOGY

D'Souza argues that the universe was created and "fine-tuned" for humans by God:

> *Fantastic though it seems, the universe is fine-tuned for human habitation. We live in a kind of Goldilocks universe in which the conditions are "just right" for life to emerge and thrive.*
>
> *The latest discoveries of modern science support the Christian claim that there is a divine being who created the universe.*

The latest scientific discoveries argue otherwise. The fine-tuning argument is rejected by scientists as just another "designed by an intelligent designer (God)" argument without foundation. In understanding the creation of the universe, scientists have data reaching to about four hundred thousand years after the Big Bang. So far the Standard Model of physics seems to work explaining what is observed in Nature, and so far God has not been needed to provide an understanding of what astronomers observe. There is still much to learn, and as physicists get even closer to the Big Bang, new physics may be needed, for the Big Bang is a mystery. It is an exciting time in physics, but there is not a job opening for a God on the physics team to explain things.

Here on Earth, D'Souza falsely accuses science of being hung up on a dogma of "no supernatural," such as a God being the supernatural designer in evolution.

> *Modern science was designed to exclude a designer....It doesn't matter how strong or reliable the evidence {of a supernatural designer} is; scientists, acting in their professional capacity, are obliged to ignore it.*

D'Souza has not been able to accept that there is no strong evidence for a supernatural designer or God. In fact, there is strong evidence supporting Nature as the designer of the universe and life.

The door is always open for Christians with better theories than Darwin's to explain the biological design of life. Actually Christians have put forward a number of theories, which included God's involvement in evolution, such as the well-advertised Intelligent Design theotheory. All have failed not only in the science laboratories, but in legal courts, which have ruled to prohibit religious supernatural beliefs in science classrooms. If D'Souza had read the judge's opinion in the Dover Pennsylvania case on the local school board's attempts to teach Intelligent Design in Dover's public science classrooms, he would understand why science and the government exclude it.

DARWIN'S THEORY

The most indigestible natural science advance for D'Souza and Christians in general is Darwin's Theory of natural selection, or evolution, for it addresses God's greatest creation, man, with a godless theory. Science states that Darwin's Theory is more than a theory and covers all biological life. It has been so successful technically and proven so many times in Nature and in the laboratories around the world that most Christians no longer dismiss it as "just a theory," although there are still some who doggedly continue to do so.

There is a spectrum of Christian acceptance of Darwin's Theory; from Fundamentalists with no acceptance to Progressive Christians who accept the whole theory. D'Souza is in the middle of Christian acceptance of science—a Liberal who tries to hold on to the belief that a supernatural God was man's creator and designer while trying to recognize that Darwin's theory works. Darwin's Theory is an inconvenient fact that forces D'Souza to search for an argument on which he can put a Christian spin on evolution in order have a Christian version, a theotheory, but his attempts at marrying supernatural religion with man's natural evolution fail for the same reasons as those of the other Christian authors.

NATURAL WORLD

Science, such as natural selection by Darwin's Theory, only works in the natural world, which does not include the supernatural. D'Souza's view of the world is different from that of the natural world, for he declares that the supernatural scriptures and science are not contradictory:

> In summary Christians should be suing to get Atheist interpretations of Darwin out. Through evolution, rightly understood, Christians can affirm that the book of nature and the book of scripture are in no way contradictory. In fact, both affirm the notion of a universe and its creatures that are the product of supernatural design and divine creation.

D' Souza forgets that most scientists are non-theists and support Darwin's Theory. This statement affirms D'Souza's belief that Nature is a product of his divine God, a belief that most scientists reject, for they argue that the universe and its creatures are the products of Nature.

CHANCE

Christians, argues D'Souza, do not believe that man is:

> . . . *an arbitrary product of time, chance, and natural forces, a mere grab-bag of atomic particles, a conglomeration of genetic substance.*

In contrast, Darwin's Theory argues the opposite: all life is an arbitrary product of time, chance, and natural forces. Chance is a critical part of random mutations, random environmental factors and natural selection, the process that makes evolution work.

DESIGN

D'Souza argues that the evidence of biological design by Darwin's Theory supports supernatural (intelligent) design in the evolutionary process.

> *Darwin's theory of evolution, far from undermining the evidence for supernatural design, actually strengthens it.*

This is a completely false statement, but Christians often attack the tenets of Darwin's Theory for it does not involve God. However the evidence for evolution is that Nature designs without a designer. Having Nature as the authority for evolution removes the need for a God, so Christians attack Darwin's design tenet. On biological design, D'Souza's falls back to the long–refuted arguments Rev. Paley made two hundred years ago that biological design is by God:

> *The Atheists (Dawkins, et al) have led people away from the real explanation of evolution put forth by the Christian Rev Paley two hundred years ago.*

Further:

> *But the universe that lawfully produces finches, moths, and humans is quite clearly the product of intention and creative design. So Dawkins's "refutation" of Paley fails gloriously and completely. Paley was right all along. It should be clear from all this that the problem is not with evolution. The problem is with Darwinism. Evolution is a scientific theory; Darwinism is a metaphysical stance and a political ideology.*

Paley's explanation for biological design being clearly the product of intention and creative design, or intelligent design by God, has long been rejected by the scientific community at large. It's incredible that D'Souza refuses to acknowledge one of man's greatest intellectual revolutions, the Darwinian Revolution. People have not been led away from Paley's explanation; people have been drawn to Darwin's theory, for it agrees with the data we observe in Nature, while Paley's does not.

D'Souza attempts to paint natural science being restricted by scientists so that miracles and the supernatural are excluded. They are, but D'Souza fails to understand why: godless science works, and it has worked through experimental verification, and all attempts to inject the supernatural have failed. So far science has succeeded with its secular approach, and the religious community has not given any reason for godless science to change.

MORAL SENSE

D'Souza can only see his religion as the source of morality:

> *If religion…can be systematically analyzed and explained as a product of the brain's evolution, its power as an external source of morality will be gone forever.*

Evolutionary biologist E. O. Wilson and others argue that the mind itself is the product of evolution and free moral choice is an illusion.

> *We can be proud as a species because, having discovered that we are alone, we owe the gods very little.*

Moral sense is a product of evolution, but other factors can influence one's morality. Religion can be an external source of morality that modifies our evolved moral sense.

DARWINISM

Knowing that Darwin's Theory cannot be dismissed easily, D'Souza devises a rationale that attacks it indirectly: he calls it Darwinism and defines it as a spin on Darwin's Theory by Atheists.

> *Darwinism is the Atheist spin imposed on the theory of evolution.*

Atheists and scientists use the principles of natural selection to develop expanded views of the modern evolutionary theory. In effect, Darwinism is the attempt by scientists to understand the breadth of the theory of natural selection. As with any science theory, its reach may be overstated, but science will correct such overreaches. D'Souza's statements that Darwinism is a metaphysical stance and a political ideology is overreach. Such views are out of mainstream science and do not reflect the validity of Darwin's Theory.

D'Souza's goal with this argument is to convert a sound scientific theory into an anti-religion plot by Atheists. His attempt to confuse

Darwin's Theory with an Atheistic ideology (Darwinism) is so obvious it fails when the first ray of reason falls on it.

GALILEO REVISITED

To provide an example of his second theme, D'Souza revisits the trial of Galileo by the Vatican in 1633 as an illustration of Atheist writers using a trial to overstate the evils of the church with a vengeance. Historians, both Christian and others, have amply documented Galileo's trial and that period of time in which church theocracies had the power to jail or kill persons for going beyond the church's geo-centric dogmas. In the case of Galileo, the Vatican decided that he had violated their dogma on the geo-centric theory of the universe by conducting experiments, writing about and supporting Copernicus' heliocentric theory as a replacement to the church's dogma. Such activities were considered heretical, and after study of Galileo's activities, the Vatican's Inquisition brought Galileo to trial. The Vatican was the judge, prosecutor, and jury, and they proceeded to find him guilty, not of bad science but of religious heresy. The church could have taken many actions against Galileo, already an old man, but they pressed ahead with his trial and found him guilty of being a "suspect of heresy" and put him under house arrest for rest of his life.

His book discussing Copernicus' theory was put on the Index of prohibited books for 125 years, but Galileo's work and his book had already been disseminated by the scientific community, which eagerly studied his work and Copernicus' theory. In 1992 the Vatican said oops, it had come to the wrong verdict in the Galileo trial. Surely taking 359 years to overturn the trial's verdict is a record for any bureaucracy to correct a mistake.

D'Souza takes this event and rewrites the character of Galileo to be that of a pushy old scientist and the members of Vatican's Inquisition at the trial to be reasonable and considerate clergy. With these rede-fined characters, he twists and rewrites the implications of the trial:

> *Galileo was a great scientist who had very little sense. He was right about helio-centrism, but several of his arguments and proofs were wrong. The dispute his ideas brought about was not exclu-sively between religion and science, but also between the new science and the science of the previous generation. The leading figures of the church were more circumspect about approaching the scientific issues, which were truly unsettled at the time, than the impetuous Galileo. The church should not have tried him, but his trials were conducted with considerable restraint and exemplary treatment.*

D'Souza's version has Galileo deservedly being punished and the church's scientific and moral reasoning elevated to "Well, they did what they had to do" under the circumstances. Being "more circum-spect," the Vatican wrote, is more important than getting the science right, but it's hardly an argument accepted by scientists then or today.

D'Souza is not the first Christian apologist to argue that the church acted properly punishing Galileo with the information available at the time. Cardinal Ratzinger, later Pope Benedict XVI, argued in 1990 that "[The Vatican's] verdict against Galileo was rational and just." Later in 2008 the pope cancelled a trip to Sapienza University, where his previous comment on Galileo resurfaced and was protested by scien-tists. D'Souza's apologetic review of the Galileo trial should also not go unnoticed nor unchallenged, for it is a case of rewriting history to meet one's wishes. Bad history should be exposed, as Sapienza university did.

DEMOCRACY

D'Souza goes off the deep end in his misrepresentation of who the Founding Fathers were and what they were trying to achieve in advancing the constitutional concept of the separation of church and state. Up until the American democratic experiment, theocracies that coupled religion with government were the norm in Europe. The Founding Fathers had knowledge of the religious wars in Europe as a cause for many deaths from theocracies of various denominations of Christianity and wanted to avoid this in the new country. In order to have religious freedom, it was necessary to have our new government separated from religion, for it was clear that to have freedom of religion, all religions had to be honored equally in the new country. Therefore, no one of them could be within the government and hold a special place. Religious liberty for all peoples and religions depended on separating religion from the government.

Religious freedom was not a gift from D'Souza's Christian God, as he argues. Which religion says it's fine to remove it from a position of power? The Vatican had a hard time accepting the democratic view of religious freedom, and as late as the mid-1800s, the Vatican was still arguing that the best form of government was a Catholic theocracy. The view that one should have the liberty to worship any or no Gods was a secular view they did not like. D'Souza's attempt to argue that the separation of church and state was a Christian idea fails the test of history.

> *The genius of the American founders was to go beyond tolerance to insist that the central government stay completely out of the business of theology. Despite its novelty, this idea was a profoundly Christian one; they were following Christ's rule to keep the domains of Caesar and God separate.*

It was not *beyond tolerance* that the Founding Fathers separated religion from government, but the simple fact that it is not possible to have a democracy guided by *we the people* in a theocracy guided by *I am the one and only* supernatural God. Jesus may have told his local group of Christians "render unto Caesar," but after Constantine in 350 AD brought Christianity into his government, almost all Christian governments from then to the American experiment were run as Christian theocracies.

The majority of the Founding Fathers were Christian. A few, such as Jefferson, Franklin, and Madison, were Deists. They rejected having the word "God" in the Constitution, clearly assigning the foundations of the Constitution to its secular citizens and not a God. It was a document *of the people, by the people, and for the people* (Lincoln's words), not a document gifted by God. Since *the people* were of many different religious persuasions, it had to be a secular constitution.

The founders, being mostly Christians, retained some of their Christian heritage. Even Jefferson, perhaps the least religious of them, argued that religious faith was a right of each citizen, and with Christians in the majority, many Christian trappings appeared and lingered after the founding of the new country. D'Souza notes:

> *After the Revolutionary War, the founders continued to hold public days of prayer, to appoint chaplains for Congress and the armed forces, and to promote religious values through the schools in the Northwest Territory.*

D'Souza expands these trappings of religion accepted by the new government into arguing that there was no separation of church and state. He misses the fact that many of these early trappings have disappeared; promoting religious values through the schools was replaced when support of the Northwest Territories was replaced by public schools on a national basis with the constitutional ruling that public schools cannot promote religious values, and the national day of prayer was rejected by Jefferson.

The wall separating the church and state is continually being tested. The third president, Thomas Jefferson, gave his understanding of the wall:

> *Believing with you that religion is a matter which lies solely between man and his God...I contemplate with sovereign reference that act of the whole American people which declared that their legislature should "make no law respecting an established religion, or prohibiting the free exercise thereof," thus building a wall of separation between church and state.*

Exactly where the wall is located is a continuing argument and has resulted in many lawsuits on specific events and places of religion in the public area. Even among presidents, it's not settled: Jefferson canceled the national day of prayer that Washington and Adams had allowed, while other presidents have accepted it. D'Souza argues falsely that:

> *Today courts wrongly interpret separation of church and state to mean that religion has no place in the public arena, or that morality derived from religion should not be permitted to shape our laws. Somehow freedom for religious expression has become freedom from religious expression. Secularists want to empty the public square of religion and religious-based morality so they can monopolize the shared space of society with their own views. In the process they have made religious believers into second-class citizens. This is a profound distortion of a noble idea that is also a Christian idea. The separation of the realms should not be a weapon against Christianity; rather, it is a device supplied by Christianity to promote social peace, religious freedom, and a moral community. If we recovered the concept in its true sense, our society would be much better off.*

Almost every assertion in this paragraph is incorrect. The separation of church and state is a constitutional concept adopted by the Christian majority of the Founding Fathers. First, religion has a place in the public arena, but it is limited so as to not infringe on the religious rights of others and cannot be sponsored by the government. Second, religious-based morality can be suggested by believers, but to be part of a law, it must be approved by the people, who can be of many different religions. Third, Christians are no more second-class citizens than citizens of other religions or those with no religion. And fourth, the separation of church and state is not a weapon against Christianity or any other religion, it is applied uniformly to all, for it is a necessary constitutional commandment for any democracy.

SECULAR EDUCATION

D'Souza takes a rather hard view of secular public education:

> *It seems that Atheists are not content with committing cultural suicide—they want to take your children with them. The Atheist strategy can be described in this way: let the religious people breed them, and we will educate them to despise their parents' beliefs. So the secularization of the minds of our young people is not, as many think, the inevitable consequence of learning and maturing. Rather, it is to a large degree orchestrated by teachers and professors to promote anti-religious agendas.*

This is indeed a remarkably anti-American view of public education. Americans have long been proud of the early introduction and support of secular public education at the state and national level. One may ask, "How can one educate the citizens of all religious persuasions in a democracy if not by the separation of church and state?" So it is not only Atheists who support secular education, but citizens of all religions who know that public education is secular for all. Any religious part of a child's education can be obtained through each family's church-supported schools.

Surely D'Souza is aware of the difficulties presented trying to teach subjects to secular students in a religious school. Take natural science as an example. Within every religion there is a spectrum of beliefs on science. Within Christianity there are Fundamentalist Christians who would argue that schools should teach biblical literalism, such as the Earth is seven thousand years old. Other Christians who may be scientists would disagree and argue to have the scientific age of 4.5 billion years for the Earth be taught. Christians are free to use whatever age of the Earth they believe in their churches, but not in public schools. D'Souza's attacks on science are based on scientific and constitutional ignorance. Teaching secular natural science may be a problem for some Christians, but it is not a problem for most of our public schools' children.

HISTORICAL CRIMES

D'Souza declares that a major theme of Atheist discourse is the historical crimes of religion. Crimes by Christian leaders and institutions are part of history. Killing people in the name of God is not limited to Christianity, but Christians have done their share. The Vatican's Inquisition, the Christian Crusades, the Christian religious wars, and the Christian witch trials are events that Christians led, and they resulted in the deaths of many. History also has many mass killings for which non-Christians hold the responsibility. Such killings are morally wrong whether they are performed by Christians or non-Christians. The general moral indictment of history is that men and institutions when unchecked can use their power to kill; the greater the power the greater the killings, whether the leaders are Christian or not. It is a fact that at one time in history, Christian institutions had great theocratic power and exercised that power and killed many.

D'Souza's unhappiness over discussions of Christian mass killings is misplaced. History's moral indictment of Christians, along with non-Christians, for killings is a fact that D'Souza does not want to hear, for that would force him to acknowledge that organizational morality is the same for Christians and non-Christians.

SUMMARY

D'Souza refuses to acknowledge the scientific and governance advances made since the fifteenth century, when Christendom was at its peak. Believing that a good offense is the best defense D'Souza misrepresents Darwin's Theory even knowing that it has been tested many times, and all evidence supports that it is a theory that does not require a God.

In addition to D'Souza's false assertions on science theories, he has made false assertions about Atheists, who he claims are responsible for a *war* against Christendom. These assertions are also without merit but are consistent with the arguments of other Christians who also cannot come to fully accept that science, like religion, is an independent discipline. The conspiratorial approach he uses to create fear is akin to that used by Joseph McCarthy, with the evils of communism being replaced by the evils of Atheism and secular science.

As members of one of the many religions, albeit the largest in our country, Christians must expect to live with the rough-and-tumble contest of ideas in our dynamic democracy. Knowing that there has been a decline of Christian authority over the years in laws, moral pronouncements, and science, Christians would do well to adjust to the fact that now the secular scientific community speaks for science and our secular democracy speaks for societal laws and morals.

Christianity remains recognized as one of many valuable, independent religious contributors to our marketplace of social and moral philosophies and serves as a major provider of humanitarian services.

The energy D'Souza expends as a conservative Christian salesman would have been more productive if he had honored natural science's theories, democracy's separation of church and state, and focused on championing Christianity's rightful place as one of many religions serving the humanitarian needs of our secular democracy.

Darwin's Gift-Ayala

F rancisco Ayala's book *Darwin's Gift, to Science and Religion* gives an insightful overview of Darwinian evolution, or as he prefers to say, Darwin's Theory of natural selection. Ayala gives further discussions of Darwin's Theory in a following book, *Am I a Monkey?* and in a paper, *Design without a Designer.*[86] He was the winner of the 2010 Templeton Prize, which noted that he is known for his achievements as an evolutionary geneticist and for his opposition to the entanglement of science and religion while also calling for mutual respect between the two. Ayala sees Darwin's Theory of natural selection as one of the most significant intellectual achievements in science by:

bringing the magnificent adaptations of organisms and their amazing diversity to the domain of science: explanations by natural laws and processes.

By using Darwin's Theory without qualifications, he does not need to employ a theotheory to circumvent or replace Darwin's Theory in the natural world, as the Liberal Christian authors have done. He discusses why he has rejected the theotheories of Creationism and ID offered by other Christians as replacements for Darwinian evolution.

Ayala's full acceptance of Darwin's Theory is in conflict with the other Christian authors reviewed.

> *It was Darwin's greatest accomplishment to show that the complex organization and functionality of living beings can be explained as the result of a natural process—natural selection—without any need to resort to a Creator[79] or other external agents.*

As a scientist and a Christian, Ayala places science in perspective with religion and dismisses the fears of science being materialistic. He confines science to the natural world, where it does not address the Christian narrative in the supernatural world of religion. The worlds are separate and independent.

> *People of faith need not be troubled that science is materialistic. The materialism of science asserts its limits, not its universality. The methods and scope of science remain within the world of matter. It cannot make assertions beyond that world.*

DARWIN'S THEORY

Ayala endorses Darwin's Theory as the foundational theory in biology for the evolution of all living organisms. His acceptance of Darwin's Theory is compatible with that of the scientific community at large. I use his detailed descriptions of Darwin's Theory as a reference in the chapter Nature's Way and in the critiques of the five books by Christian authors.

THEOTHEORIES

Ayala has long been a critic of the Christian theotheories that have been proposed by various Christian groups to replace Darwin's Theory in public school science classes. He faulted the Scientific Creationism theotheory as being bad science and bad religion when it was proposed for use in a public school. When Scientific Creationism was challenged in court, Ayala was an expert science witness against the Arkansas state law in the *McLean v. Arkansas* case in 1982. Later in 1987, the *Edwards v. Aguilar* case dealt with a similar law in Louisiana, and Ayala actively opposed it. In this case the Supreme Court ruled that teaching the Scientific Creationism theotheory was unconstitutional because it endorsed religion, and as such, it should not be taught in the public schools. Ayala has opposed a follow-on Christian theotheory, Intelligent Design (ID), as also being bad science and bad theology. He quotes the ruling by the presiding Judge Jones in the *Dover Area School District* case:

> *It {ID} has not generated peer reviewed publications, nor has it been the subject of testing and research. ID is not science and cannot be adjudged a valid, accepted scientific Theory. {It is an} alternative masquerading as a scientific Theory.*

ID proponents argued that Darwin's Theory is anti-theistic, that is, it does not support a belief in the existence of a supreme being. But this attack also failed in the eyes of Judge Jones, who used an argument by the National Academy of Sciences that natural science does not deny the existence of a divine creator God, it simply does not address the supernatural. Judge Jones gave the judgment against ID after the best minds supporting ID (Michael Behe, et al.) came to its defense; their arguments failed.

SCIENCE AND RELIGION

Ayala, being both an evolutionary biologist as well as a Christian, addresses both the natural world of science and the supernatural world of his Christian religion. On science:

Science is a wondrously successful way of knowing the world.

On the range of science:

Science transcends cultural, political and religious beliefs because it has nothing to say about these subjects. That science is not constrained by cultural or religious differences is one of its great virtues. It does not transcend these differences by denying them or taking one position rather than another. It transcends cultural, political and religious convictions because these matters are none of its business.

And on religion:

Religion concerns the meaning and purpose of the world and human life, the proper relation of people to their Creator and to each other, the moral values that inspire and govern their lives.

Ayala has defined separate concerns for Nature's science and God's religion. He then argues that these different worlds are compatible:

Science and religion are compatible because they concern different domains of knowledge.

Science and religious beliefs need not be in contradiction. If they are properly understood, they cannot be in contradiction because science and religion concern different matters. Scientific conclusions and religious beliefs concern different sorts of issues, belonging to different domains of knowledge; they do not stand in contradiction.

As Ayala noted, science and religion each has a realm or domain of knowledge that differs fundamentally—knowledge in the natural world is verified by the scientific process, while knowledge in the supernatural world does not require scientific verification. Science does not contradict religious issues, for it just does not address them. For example, take the issue at hand, the question of *who to thank for evolution*. If answered using the natural world's scientific knowledge, the answer is Nature; if answered using the knowledge of the religious supernatural world based on the scriptures, the answer is God. Science is not concerned with answers from the Christian or any god as long as such answers remain separate from science.

Why should totally different worlds, or domains, of knowledge be compatible? We know that Christian knowledge varies between denominations; example, Catholic knowledge is different from Methodist knowledge, which is different from other Christian denominational knowledge, and in some cases the knowledge can be incompatible. With the complexity and diversity of our world, would it not be beneficial for man to have several different views, various religious ones from belief and one from scientifically verified information in the natural world totally separate from any religious variation to ponder?

Ayala uses "theological parlance" to explain God's involvement in the creation of the universe In theological parlance, believers argue that God may act through secondary causes, a critical step for Christians believing that God created the world. "Theological parlance" is another way in this case to say supernatural belief of God relying on Nature's science.

Finally, Ayala outlines how some theologians and other people of faith would like to see evolution:

> *Many theologians and other people of faith see evolution as the process by which God creates the wonderful diversity of the living world. Thus I would add, paraphrasing theologian Aubrey Moore, that evolution is not the enemy of religion, but rather, its friend.*

If theologians see the theory of evolution as a friend in their supernatural world, that is fine, for they can describe it as they wish. But theologians must realize that non-theists in the natural world see evolution as a natural process operating with Nature's tool set (physics, chemistry, and biology) without God's actions to produce man and the diversity of life. Nature's evolution does not differentiate between enemies and friends. Christians may see evolution as a friend, but in the natural world, Mother Nature doesn't care; she just does her thing. She is never a friend, nor an enemy; she just is.

Ayala's discussion of values brings up a need for definitions:

> So these {science and religion} are different matters. Science has no way of exploring values.

Science does not attempt to explain personal values for these are a product of the individual. However, the utility of values by individuals and groups are explored by science. Sam Harris' recent book, *The Moral Landscape: How Science Can Determine Human Values*, gives a glimpse into some of the neuroscience research on values. Christians may find their religion of great help in assembling or modifying their values, but science can be helpful studying an individual's values in context with those of groups of interest. Having science engaged does not minimize the importance of morals and values that are received by individuals from their religion. A person's values are an individual responsibility, as is searching for the meaning and purpose of life. The biblical narrative can provide guidance on these for Christians, but there is other guidance available.

Ayala's discussion of religion and science is his answer to the question, "Can a Christian be a Darwinian?" and the answer is yes, for he separates and honors each individually. Many philosophers have struggled with an answer to this question, and most fail, for they do not make a clear separation between the authority of God and that of Nature. In many cases the definition of a Darwinian is faulty, for it does not meet the rigorous demands of science, which separates the scientist from the believer and does allow building bridges or mixing of the two, both of which are scientifically unacceptable acts. On the other hand, Ayala's discussion gives a more direct and helpful explanation of being a Darwinian and a Christian: one who honors both the independence and separation of the two disciplines.

THE GIFT

The gift from science to religion that Ayala sees is to ascribe the suffering of man and animal observed in the natural world to Nature and not to God. Natural evolution as outlined in a letter (from Joseph Hooker) to Darwin describes Nature's selection's process as one that "abounds in cruel behavior and is clumsy, wasteful, blundering low and horridly cruel works of Nature." Ayala quotes the philosopher David Hume, who lived a hundred years before Darwin, on this point:

> Is he {God} willing to prevent evil, but not able? Then he is impotent. Is he able, but not willing? Then he is malevolent. Is he both able and willing? Whence then evil?

The conflict with explaining "evil" in God's world is a central problem recognized by Christians, but a non-problem for Naturalists after Darwin simply because Nature's killings, vividly described by Hooker, are just part of Nature's natural selection process; it is how Nature works in the natural world. Ayala as a Progressive Christian accepts natural selection; therefore, the killings can be attributed to Nature's processes and not to God's actions.

Darwin's Theory may be seen as a friendly 'Gift' to Christians by providing answers to the dilemma of evil, but it is a 'Gift' with the strings of Nature's authority attached, which may not be acceptable to many Christians. But it is their choice. Progressive Christians can accept Nature's authority over natural processes, while Liberal and Fundamentalist Christians cannot, and for them the dilemma of evil remains.

SUMMARY

Ayala has been a champion for teaching the importance and fundamental nature of Darwin's Theory of natural selection. He summarizes his insight about the theory:

> *The theory of evolution conveys chance and necessity jointly enmeshed in the stuff of life, randomness and determinism interlocked in a natural process that has sprouted the most complex, diverse, and beautiful entities in the universe: the organisms that populate the earth, including humans who think and love, endowed with free will and creative powers, and able to analyze the process of evolution itself that brought them into existence.*

This paragraph should be obligatory reading for all Christian authors discussing evolution and particularly those trying to employ their theotheories, such as Intelligent Design, Sacred Intelligent Design, and Theistic Evolution, as replacements for Darwin's Theory. Ayala firmly describes the scientific community's position, namely, that Darwinian evolution, thank you, works quite well as Nature has given it to us.

The difference between Ayala and the other Christian authors who were reviewed on evolution is that in the natural world, as a Darwinian scientist, he uses natural science without religious involvement and does not consider the need for a religious theotheory as a replacement for Darwin.

Science asks for nothing more of its scientists than a full acceptance and appreciation of natural science, and Ayala does this unflinchingly. As a Christian, he keeps his belief in the Christian supernatural narrative as his own private matter. Having respect for the independence of both Nature's world and God's supernatural world grants him the position of being both a Darwinian and a Progressive Christian, and as such, he is comfortable thanking Nature and separately God, each in their separate worlds, for their views of man's creation and evolution.

Partnership

It is obvious that believers and followers of both God and Nature are alive and kicking in our world today with supporters on both sides of the question *who to thank for evolution—God or Nature?* With both sides likely to be around for a long time within our democracy, should we not search for a relationship, even a partnership that will, if we are wise, accommodate answers for those believing in God, honoring Nature, and respecting the State. Such a partnership would represent "symbiotic mutualism," for the organisms would be independent with an increased survival potential by having mutually beneficial relationships.

One relationship is the constitutional amendment on the independence of government from religion which follows the dictates of our Founding Fathers. Our Constitution included the principle of keeping the government separate and independent from religions. At that time no such agreement was needed for separating religions from science, for major conflicts over scientific advances had not yet occurred in America.

But since that time, conflicts have occurred, and scientific organizations such as the National Academy of Sciences have addressed the need for separating science and religion, in essence, establishing a second relationship:

> *But science and religion occupy two separate realms of human experience. Demanding that they be combined detracts from the glory of each.*

The third relationship has been the long standing understanding by the government to use the best independent natural science available for their operation and not to include religious tainted science nor the government.

History reveals that science, religion, and the state have had independent working relationships, albeit informally and imperfectly, for hundreds of years. Such relationships have informally adhered to two rules—first, each of the three disciplines recognizes that it is separate and independent, and second, representative members are the authority for answering questions within their discipline: representatives of Christian churches for religious questions, scientists from the scientific community for scientific questions, and citizens for governing questions.

An early example of a default relationship began after the Copernican scientific revolution led to the demise of the Vatican's (God's) geo-centric theotheory of planetary motion and its replacement by Nature's helio-centric theory proposed by scientists. God's authority over the science of the cosmos was replaced by Nature's authority by astronomers simply because it worked better at describing Nature. The Vatican slowly stopped defending its geo-centric theotheory dogma even though conflicting commentary remained in the Christian literature. In the natural world, the helio-centric theory gained increasing scientific support, and when widely verified by observations, it moved into the universities, then into the mainstream of public education and in everyday life for it provided such things as better calendars for the government to set holidays. The church no longer attempts to teach the geo-centric theotheory as natural science.

The removal of the church's geo-centric dogma had little effect on Christianity, the Bible, or its many believers. The Bible was not revised, believers continued to go to church, the sky did not fall on science classrooms, and further explanations of planetary motion were left to the scientific community which later included Newton and Einstein.

Much, much later the US space program, a State venture, used Nature's physics theories outlined by Copernicus, Newton, and Einstein for calculating the flights of its spacecraft to explore the moon and planets with navigational accuracy not possible with the church's old geo-centered theotheory. Christian astronauts venturing to the moon surely would give thanks for the successful navigation of their trip to Nature's godless theories of planetary motion.

A second scientific example is the Darwinian Revolution. Although widely accepted by scientists as the basis of our success-ful understanding of evolutionary biology and employed by many businessmen in the successful bio-technology industry, its acceptance by many Christians has been slow. High religious emotions over accepting Nature's Darwinian evolution to describe man's creation, design, and evolution have presented barriers for acceptance for many Christians. Polls, although noting a ninety percent acceptance by evo-lutionary biologists, indicate that two-thirds of Christian Americans want a supernatural God dictating creation-evolution, not Nature's science. Fortunately this view is essentially ignored by bio-research centers, public universities, and government laboratories trying to solve biology problems in today's world.

Present misunderstandings of who to thank for natural science answers, God or Nature, are a reflection that public school science programs have failed to give students, or at least enough of them, an adequate education in science and an understanding of why science should be the authority in the natural world. Also, Christian churches have failed to educate their believers that God's authority in his supernatural world will not be lost if Nature is given the authority over science, yet it is widely known that scientific advances utilizing Darwin's Theory have led to medicines and treatments that have saved the lives of many Christians and non-Christians alike. Hopefully the reality of the value of scientific advances to mankind will be recognized and lead more Christians to accept a partnership with science.

Another example comes from the Democratic Revolution, which produced our secular Constitution that forbids the establishment of religion, including Christianity, by the government. Freedom of religion is guaranteed by the Constitution for all religions, but they are to be separate from the government. The authority of secular constitutional law, not biblical law, is used for governance throughout the country.

Although periodically challenged by some Christians, the Constitution has served to uphold the freedom of religions, as well as no-religion, for all citizens. Importantly, it has ensured the freedom of all the religions, including the minority religions. Most Christians have accepted the secular laws of the Constitution, but there are those who refuse to differentiate between a secular democracy and a Christian theocracy and continue to argue for the use of biblical laws in government.

Another example comes from the Neuronian Revolution, which is increasing our knowledge of the biological evolution of man's brain, his moral sense, and his mental capabilities. A deeper understanding of man's mental capabilities by neuroscientists is helping to solve mental problems affecting peoples of all religions. As benefits of neuroscience research unfold and become recognized by the public it is hoped that there will be a greater acceptance of the fruits of the Neuronian Revolution and an increased support for a partnership of religion, science, and state.

It is also hoped that Christian scholars will expand their arguments that the Bible is not a book of science, for its writers did not intend it to be one. A broad array of church leaders, from St. Augustine to Pope John Paul II, as well as Christian scientists as far back as Galileo and Kepler have noted that personal salvation, the human condition, and humanitarian activities are the mission of the church and not the teaching of science.

There are Progressive Christian leaders today who are providing insights into new and acceptable Christian positions on science, as Bishop Spong notes:

> So wrestle with {read} Dawkins, but recognize that there are people in the Christian Church who can read and even appreciate Dawkins and still be drawn into the worship of God and a Christian understanding of life.

In summary, there is a wide range of beliefs held by Christians on their acceptance of scientific advances. Progress toward a partnership with science and the state has begun, albeit limited, with an increasing number of Progressive Christians joining with Naturalists.

Further progress on expanding the partnership will depend on other Christians joining. It will require Liberal Christians to recognize the scientific sin Richard Feynman warned about—it is easy to fool yourself—there are no special Christian theotheories to replace science theories, there are only natural science ones. It will require recognizing that there are fundamental differences between understanding God and Nature and that these differences should not be papered-over with unsupportable theotheories that attempt "marriages" or the construction of "bridges" with bad theology and bad science.

It is clear that an answer given in the natural world to a conflicting issue between the science and religion cannot be both theologically and scientifically sound, for any one answer cannot be both under the authority of a supernatural God outside of the natural world and under godless Nature in the natural world. Their authorities are mutually incompatible; answers are either under the authority of God or Nature, never both jointly. The separation and independence offered by a partnership is the answer to giving Nature, God, and the State respect and the authority over their respective disciplines.

The Nature-centric processes of the Darwinian and Neuronian Revolutions describe the evolution of man and his mental capabilities, which have allowed him to invent religious narratives, supernatural gods, and religions. In short, those who accept the science, Naturalists and Progressive Christians, can declare:

Thank Evolution for God.

Since the evolution of living organisms is a secular natural process, it can be used by all peoples of all religions or no religions who honor the independence of Nature and its handmaiden, science. For Christians, their position is dependent on the degree of independence they give natural science from their beliefs of their supernatural God.

For Progressive Christians, it is through their belief in the narrative of their God in his supernatural world and separately in their acceptance of Nature in the natural world that gives them a dual view. This dual belief results in both a Nature-centric answer for the natural world and a God-centric answer for supernatural narratives, or, in short:

Thank Evolution (in Nature's world) for God and

Thank God (in God's supernatural world) for Creation.

This allows Progressive Christians to join with Naturalists in a partnership based on the separation and independence of God.

However, there are other Christians, Liberals, who do not totally accept Nature's authority over evolution in the natural world and, accordingly, reject *thank evolution for God* and replace this Nature-centric theory of evolution with a God-centric theotheory that modifies Darwin's Theory. This belief leaves Liberal Christians with continuing conflicts with natural science, for their God-centric argument erroneous mixes religion and science when they declare:

Thank God for Evolution.

The scientific community does not accept God-centric theotheories, so this answer only perpetuates the conflicts between religion and science for it is built on bad Christian theotheory and bad science.

A third group of Christians, the Fundamentalists, do not accept Darwin's Theory of evolution and insist on a literal reading of the biblical Genesis, so their answer is based on the God-centric belief of the biblical creation narrative, so they declare:

Thank God for Creation

This is purely a religious belief that does not attempt to rewrite Darwin's Theory. It is an answer that rejects natural science and affirms their answers from the supernatural world of religion. It does prevent their involvement in a partnership and continues their conflicts with natural science.

Thus, an answers to the question, *who to thank for evolution?*, reflects each individual's belief in God's extent of authority. But since we live with religious freedom provided by our Constitution conflicts are allowed, but tolerance is necessary. With a little unauthorized editing, we can use Thomas Jefferson's[87] observation on religious tolerance to save the day:

> *But it does me no injury for my neighbor to say there are twenty gods or no god {in the supernatural world}. It neither picks my pocket nor breaks my leg {in the natural world}.*

Progressive Christians and Naturalists are already partnership members. If Liberal Christians wish to be in the Partnership they can be assured that neither will their pocket be picked, their leg broken, nor their belief tossed out if they accept membership into a partnership with natural science and democracy.

Fundamental Christians have little use for science and are not expected to be likely partnership candidates, as they have concluded that understanding Nature is not important.

Finally, it is important to note that there is urgency to supporting the partnership in order to address ongoing human miseries resulting from roadblocks erected by current conflicts between religion, science, and the state. At the top of the list of critical problems is the population problem—too many people for what societies, even the planet, can support today. A fourth of the people in the world are undernourished, and a comparable number are dying for the lack of the basics—food, water, and medical services. Christians engaged in charitable activities know that there is much humanitarian work crying to be done. The situation worsens as more and more children are born each year with the tacit approval of most religions and states. How does arguing with science help Christians deliver humanitarian services, particularly since Christians have lost every battle that pitted biblical science against natural science? What a waste of Christian energy.

Science is sounding alarms for the future health of the planet, from global climate change caused by atmosphere pollution to the unsustainably large population and their profligate use of polluting energy sources. Efforts to solve these human problems without a partnership between science, religion, and state have failed for far too long.

To address these pressing problems, religions and governments are urged to carry science's message to the people and see that public education is producing students knowledgeable with Nature along with their God. The Earth is man's home in Nature, and its mistreatment by a population too large, too invasive, too uninformed of science, and too insensitive to pollution, will leads to fatal outcomes for many. Mother Nature can only be expected to do her thing, and her solutions for overpopulation are not expected to be pretty. The facts from science, the compassion from religions, and the actions by governments are vitally needed to accomplish meaningful humanitarian services that may hopefully address Mother Nature's actions favorably in the future.

Epilogue

A s an optimist I believe that after another 2,500 years, Socrates will continue to be a relevant philosopher, and his followers will continue to stress the importance of questioning and challenging issues. A partnership between religion, science, and the state will be accepted by many, unappreciated by some, and contested by others. Only by questioning will the critical need for a partnership of science, religion, and the state continue to be understood and given the chance to prosper in the ever-contentious world of humans for there is a challenge in every new generation to educate the need for separating religion from science and governance from the relentless pressure of religious proselytizing.

At universities, the new Socrates T-shirt of the day will proclaim:

5,000 years and still corrupting the youth

Departmental sponsors:

Philosophy, Physics, Chemistry, Biology, Neuroscience, Social Science, Engineering, Literature and Religious Studies.

Appendix A: Modes of Interactions

Comments on Modes of Interactions
between Science and Religion

T he author Eugenie Scott, Executive Director for the National Center for Science Education, in her book *Evolution vs. Creationism,* outlines four commonly used modes of interaction in discussions between science and religion: Conflict, Dialogue, Integration, and Independence. I selected the fourth mode, Independence, to answer the question addressed in this book: *"who do I thank for evolution—God or Nature?"* Comments on the utility of using each mode are summarized below.

CONFLICT

Statement—*Both religion and science regard the other as an enemy.*

Comment—There are thousands of religions, each with a unique supernatural narrative. Science does not address the supernatural god or narrative of any religion, so it does not conflict with any religion or regard any of them as an enemy. Science and religion do conflict when a religion steps onto science's turf. For example, scientific measurements have determined the age of the earth to be 4.5 billion years. This is a conflict with the age of seven thousand years, which Fundamentalist Christians have attempted to teach in public school science classes. The thousand religions may use any date they wish in their supernatural narratives as long as they stay in their supernatural world and off science's turf in the natural world.

Conclusion—Conflict interactions are not productive and should be avoided by keeping science and religion separate and independent.

DIALOGUE

Statement—*Dialogues may arise when religions have difficulty answering questions that religions cannot answer. Example: why is the universe orderly and intelligible?*

Comment—Science cannot answer this science question at this time, so there is reason for a dialogue with everybody, including religions with miracles. Any suggestion of injecting miracles will end any discussion. Scientists usually respond to questions that they cannot answer at the moment by working until they find an answer.

It is important for science not to address religious positions that inject religious supernatural miracles into science. Each should be respected for what each is and what it says and the fact that each is independent. Science is a search for understanding Nature in the natural world, no miracles allowed. Religion may employ miracles but has inadequate discipline to engage in the search for natural science answers.

Conclusion—Dialogues between science and religion may be interesting, but they serve little purpose in helping believers improve their religion or scientists advance science. Supernatural religion is never an answer to scientific questions.

INTEGRATION

Statement—*Christian scientists pursue issues that they declare are both theologically and scientific sound.*

Comment—There are no issues that are both theologically and scientifically sound, so little comes from such integration discussions.

Statement—*Theologians have pursued an approach, theology of nature, in attempts to find a proof in nature for the existence of God, such as the universe being fine-tuned by an intelligent designer so that it supports life.*

Comment—Theology of nature, or natural theology, means theology supported by science and not dependent on the miraculous. Some philosophers have proposed and endorsed this theology, and although much has been written, it has yet to produce a common philosophical framework with which to integrate science and religion. Christians often cite the example of a fine-tuned universe to support natural theology, but this example is just the discredited Intelligent Design theotheory applied to the universe. ID has already failed when applied to biology of man, and it fails when applied to the physics of the universe. It must be rejected, for Nature holds that the universe was created without a creator and designed without a designer by its tools, physics, chemistry, and biology, no god required. What is there about Nature's physics, chemistry, and biology that theologians do not understand that makes them avoid Nature's answers?

Conclusion—Few useful results have been forthcoming from attempted integrations of religion with science simply because any religious supernatural integration into natural science automatically converts the science theory into a religious theotheory, reducing the discussion to believers talking to believers.

Science and religion must be separate and independent to exist and produce meaningful results. It must be remembered that there are thousands of religions and gods. How can all of these hundreds of religions be integrated with the one science when they cannot be integrated among themselves?

INDEPENDENCE

<u>Statement</u>—*Religion and science assertions are in two different languages that do not compete because they serve contrasting questions.*

<u>Comment</u>—This is not true, for each should be unfettered to ask or answer any question that may come its way. For example, "Who was the creator of the universe" is a question that both religion and science should be asked. The fact that the answers, if any are found, will differ is to be expected, for their knowledge bases and the processes used by each are so very different.

<u>Statement</u>—*Religion and science offer complementary perspectives on the world that are not mutually exclusive.*

<u>Comment</u>—Perspectives from science and religion may be in conflict and non-complementary, for their authorities are always in conflict. Having them give complementary perspectives on any issue would be only an accident.

<u>Statement</u>—*We can accept both science and religion if we keep them in separate watertight compartments all of our lives.*

<u>Comment</u>—True. Science relates to the natural world and religion to the supernatural world, and they should be separated into separate watertight compartments—that is, keeping the natural world separate from the supernatural is the basis for removing conflicts.

<u>Conclusion</u>—Both religion and science are important and vital disciplines. Each should respect the authority and independence of the other—religion describing god's supernatural world and science describing nature's world. The thousands of religions need to be separated from the one science. It is only by separating

the many religions from science and assigning the question to either the supernatural world of God or the natural world of Nature that the question of *who should we thank for evolution?* can be addressed.

Appendix B: Survey of Evolutionary Biologists

Extracted from *Evolution, Monism, Atheism, and the Naturalist Worldview*

By Greg Graffin

The following findings are from Greg Graffin's survey sent to 291 evolutionary biologists who are members of the national academies of science in twenty-two countries. The return rate was 54.9 percent. Five key questions in the survey and their responses are summarized below.

Question 1—Religious or Not—"Do you consider yourself a religious person?" Survey—83 percent responded—No

Question 2—Religious or Not—"Which best describes your religion?" Survey—87 percent responded—No

Question 3—Belief System—"Which best describes your belief system?" Survey—41 percent responded—Atheist 22 percent, Naturalist 16 percent, and Agnostic 3 percent

Question 4—Belief in god—"I don't believe in god (traditional definition)" Survey—78 percent responded—I don't believe in god

Question 5—Evidence for god? Survey—70 percent responded—No grounds for belief in god

A more recent survey on scientists and their views conducted by the Pew Research Center in June 2009 shows:

Scientist by party affiliation and ideology:

By party – Republican – 6%, Independent - 32%, Democratic – 55%,

By ideology - Conservative – 9%, Moderate - 35%, Liberal – 52 %,

Appendix C: Key Tenets of Darwin's Theory

Summary Reference for Biology Students

The theory of natural selection by Charles Darwin is the fundamental basis for our knowledge of the biology of all life. A leading evolutionary biologist, Francisco Ayala, has outlined the breadth and depth of the theory:

> *The theory of evolution conveys chance and necessity jointly enmeshed in the stuff of life, randomness and determinism interlocked in a natural process that has sprouted the most complex, diverse, and beautiful entities in the universe: the organisms that populate the earth, including humans who think and love, endowed with free will and creative powers, and able to analyze the process of evolution itself that brought them into existence.*

The following is a list of six key tenets from Charles

Six Key Tenets of Darwin's Theory	
1. Natural process	Natural selection is a process occurring in Nature without supernatural influences
2. Long time frame	Evolution of life on Earth, including man, has occurred over 3.5 billion years.
3. Common ancestor	All of life has evolved from a common ancestor from which the great diversity of life of today has emerged.
4. Design by Nature	All living creatures were designed without a designer as result of Nature's process of natural selection
5. Random mutations	Chance is essential to the process of evolution by natural selection, which employs random mutations.
6. Chance outcomes	Evolution is the results of random mutations within random environments

Darwin's Theory of natural selection. The outcomes include a living organism, man, with a moral sense and mental capabilities for creating supernatural narratives.

Appendix D: Overview of the Evolution of Blood Clotting

∞

EXTRACTED FROM THE EVOLUTION OF VERTEBRATE BLOOD CLOTTING
Dr. Russell Doolittle, UCSD

O nce life evolved there was a continuing demand on each system of the organism to adapt to the increasing performance demands from new, more active and complicated life forms. One such system, blood clotting, evolved to staunch the loss of blood in case of leaks, which are bound to happen when circulating fluid systems are incorporated into living organisms. This system has evolved and met the demands of new species over the last five hundred million years of evolution. For humans Doolittle notes:

> *Blood coagulation in humans is a delicately balanced process*
> *involving more than two dozen extracellular proteins, many of*
> *which need to be converted from precursor forms during the process.*

Figure E-1 illustrates the evolutionary tree of life for the sequencing, appearance and disappearance of some of the key blood clotting factors over 500 million years, from early life forms, amphioxus, to humans. During the evolution of vertebrates, there was an increasing demand on the clotting system as the pressure and flow rate of the blood increased. Not only was there a need to stop a leak, but to localize it at the site of injury, do it quickly, and construct a seal that would keep out harmful microbes. New blood clotting factors appeared during the evolution of vertebrates, and in some cases, there were factors lost for some species, as was the case of factor XII for birds some three hundred million years ago.

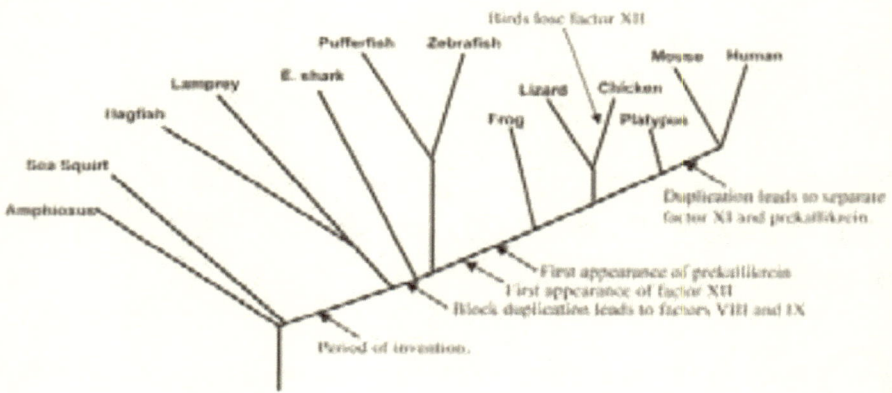

Figure E-1: Relationships of representative pre-vertebrates and vertebrates blood clotting factors along a tree of life showing approximate dates over a period of about 500 million years.

Evolution by natural selection of the blood clotting processes has been able to meet the evolving design requirements of many different species, one species at the time, over the five hundred million years of the evolution of living organisms.

The complexity, both in the number of steps and in the number of clotting factors and their interrelationship involved in the process is illustrated in Figure E-2 for various clotting factors. Each of the many steps must work in a specific sequence for the clotting to occur properly. Details of the links and the changes in the DNA, RNA, and proteins of evolving clotting processes for various key species can be found in the referenced paper by Doolittle.

The important summary observation is that the complex blood clotting process we see in humans today evolved step-by-step from earlier processes by natural selection over several hundreds of millions of years to meet the differing demands of each new species in the tree of life. By this process a blood clotting system was designed without a designer for each new species, of which there were thousands in the tree of human evolution.

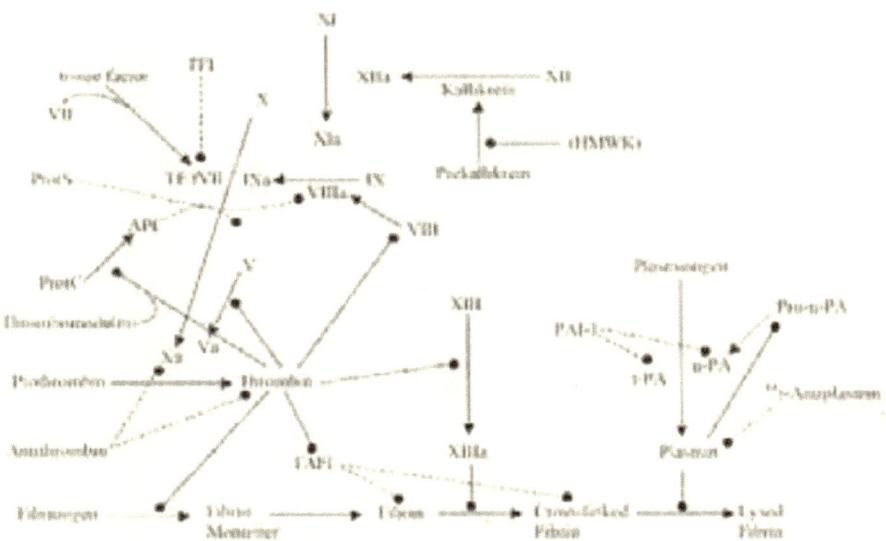

Endnotes

1   A theory may include both supernatural theological arguments and natural arguments. The combining of these two worlds of knowledge produces religious theotheories
that are not testable by science. Supernatural is used to denote God's world where information is not required to pass the scientific test and may include unverified information and
miracles beyond nature—thus supernatural.

2   Liberal Christians acknowledge evolution, but do not accept Darwin's Theory in its
fullness and invent theotheories as substitutes.

3   Nature's world, or the natural world, is described by a body of information which
includes science theories of Nature verified by the scientific process and excludes all
supernatural information. God's world is defined by the supernatural Christian narrative
describing God.

4   A community of men and women using the scientific process to understand Nature.
One example is the National Academy of Science.

5   A distinction is made between the natural world of nature, which is understood
by observations and theories adhering to natural science's methodology, and God's
supernatural world, which is understood by belief in personal revelations and scriptural
descriptions. Christians, like everyone, can access the natural world's information and
theories, but they may not change it without using natural science methodology.

6   Awarded the Templeton Prize in 2010.

7   What Is Naturalism?, Timothy Williamson, New York Times, 9/4/2011.

8   Definitions of Mythos and Logos follow from the thoughts of Karen Armstrong in the *Evolution of God*.

9   *Knocking on Heaven's Door*, Lisa Randall

10 There are exceptions, such as "marriage," child adoption, and financial arrangements.

11 *The Birth of Religion, The World's First Temple*, National Geographic, June 2011.

12 These findings suggest that religion evolved from pre-existing cognitive functions, and that it may have been subject to selection of adaptively designed system for solving the problem of cooperation.

13 *A History of Israel*, John Bright.

14 Believed by Christians to be the Christ.

15 Excellent discussion in Abraham's Progeny, and Their Texts, Edward Rothstein, October 22, 2010.

16 Hawking has stated that from the physics we know, there is no need for a god to explain the how and why of the creation of the universe. *What Stephen Hawking Has to Tell Us about the Existence of God*. Sean Carroll, Science + Religion, September 3, 2010.

17 Scientific community; composed of the People, by the People and for the People (People are constructs of nature).

18 *Origin of Eukaryotic Cells,* Lynn Margulis, 1970.

19 *On the Origin of Species by Means of Natural Selection, or the Preservation of Favoured Races in the Struggle for Life* is the full title of the first publication. For the sixth edition of 1872, the title was changed to *The Origin of Species*.

20 Named by Neil Shubin, who discovered the fossil in 2004.

21 Christopher Will's book, *the Darwinian Tourist*, presents an interesting insight into the Hobbits.

22 Einstein was referring not to the Christian God, but to Nature as the highest authority.

23 *Jefferson's Scissors*, Louis Perry, Appendix B: *"God and Morality: Is There Any Relationship between God and Morality?"* by Scott Hestevold.

24 Letter to Asa Gray, 1860.

25 The Smithsonian Museum has listed multiple hominini, including Homo floresiensis, Homo neanderthalensis, Homo heidelbergensis, Homo erectus, Paranthropus boisei, Paranthropus robustus, Australopithecus afarensis, and Sahelanthropus tchadensis.

26 *How Evolution Explains Altruism*, Orem Harman, NYTimes, April 8, 2010.

27 *Super Cooperators*, Martin Nowak.

28 Many survival skills, social skills, and morals have been left to us.

29 *Braintrust: What Neuroscience Tells Us about Morality*, Patricia Churchland.

30 *Morals for Animals, Science and Religion Today*, Tom Oord November 24, 2009.

31 *Genes Play Major Role in Primate Social Behavior, Study Finds*, Nicholas Wade, NYTimes, Dec 19, 2011

32 Russell Doolittle has outlined the step-by-step evolution of oxytocin and vasopressin in the evolution of posterior pituitary non-peptides from the lamprey to mammals.

33 *The Mirror Neuron Revolution: Explaining What Makes Humans Social*, Marco Iacoboni Scientific American July, 2008.

34 Joshua Greene, psychologist at Harvard University.

35 *The Evolution of Morality*, Richard Joyce,

36 The Moral Life of Babies, Paul Bloom, *New York Times* Magazine, May 9, 2010.

37 *Thirst for Fairness May Have Helped Us Survive*, Natalie Angier, *New York Times*, July 4, 2011.

38 *Morals Without God?*, Frans De Waal, *New York Times,* October 17, 2010.

39 *Comment,* William Keith, NYT, July 18, 2010.

40 *Are You There God? It's Me, Brain: How Our Innate Theory of Mind Gives Rise to the Divine Creator*, Jesse Bering

41 Others have proposed this view. See Graffin's survey in Appendix A. The survey shows that 71 percent of biologists believe religion is a social phenomenon that has evolved with the biological evolution of H. sapiens.

42 Hinduism is an old, vastly complicated religion with no central god(s), but one with many facets of belief and tolerance for all.

43 The last heretic burned at the stake by the Vatican was in 1826.

44 Baron Woolf, Lord Chief Justice of England and Wales, 2005, Wikipedia.

45 *Rupture with Vatican Reveals a Changed Ireland*, Sarah Lyall, *New York Times* September 17, 2011.

46 *The End of Awe*, Maureen Dowd, July 23, 2011.

47 Timothy Keece quoting 1 John 2:22.

48 Two-thirds (67 percent in Britain and 66 percent in the United States) think religious leaders should not try to influence government decision making. British Social Attitudes Survey, The 26th report, 2009.

49 *Jefferson's Scissors: Solving the Conflicts of Religion with Science, and Democracy,* L. W. Perry, addresses this struggle.

50 Letter by Abner Kneeland to *Universalist*, Thomas Whittemore, editor, December 20, 1833, published by Abner Kneeland in *The Investigator*, compiled by Annie Laurie Gaylor.

51 *Who Wrote the Bible?,* Richard Elliott Freidman.

52 Harris Poll, 2009.

53 *Tussling Over Jesus*, Nicholas Kristof, *New York Times,* January 26, 2011.

54 *A History of Israel,* John Bright.

55 *Re-Claiming the Bible for a Non-Religious World*, John Shelby Spong.

56 Two excellent books on the writing of the Bible are Richard Elliott Friedman's *Who Wrote the Bible?* and David Rosenberg and Harold Blooms' *The Book of J.*

57 An excellent summary of the rise of the Dualist beliefs of Christians and the Monist beliefs of Naturalists (Agnostics, Atheists, and Naturalist) is given in Greg Graffin's book with a survey, *Evolution, Monism, Atheism and the Naturalist Worldview*, (see Appendix A for summary of the survey).

58 *Catholicism and Science*, Hess and Allen.

59 Winner of the Templeton Prize for reconciling religion and science in 2007.

60 *Rome Fiddles, We Burn*, Maureen Dowd, NYT, July 16, 2010.

61 On Evolution, Biology Teachers Stray from Lesson Plan, *New York Times*, February 7, 2011.

62 An excellent summary of the evolution of the early theotheories is given in Eugenie Scott's *Evolution vs. Creationism.*

63 An example message by Christian college on science and religion. This one is from Bob Jones University, 2004.

64 *Saving Darwin,* by Giberson has an insightful discussion of the Scopes trial.

65 A detailed overview and history of theotheories is given by Eugenie Scott in her book *Evolution vs. Creationism.*

66 *Edwards v. Aguillard*, Supreme Court, 1987.

67 *Jefferson's Scissors, Solutions to the Conflicts of Religion and Science,* Louis Perry.

68 *The Lecture Tour of Germany, Part II: Gottingen*, Bishop Spong, April 11, 2011.

69 *Fact and Faith*, J.B.S. Haldane, 1934.

70 The scientific definition of a living organism is one that is able to acquire and use energy, has a membrane separating it from its surroundings, and can reproduce on its own.

71 *Jefferson's Scissor's* describes Bryan's intended program.

72 Bryan died shortly after the Scopes trial, so the national program was not pursued afterward.

73 *Is Nothing Sacred?*, Salman Rushdie, 1990.

74 *The Polkinghorne Reader* (edited by Thomas Jay Ord) provides key excerpts from Polkinghorne's books.

75 See Appendix A, Survey by Graffin; 70 percent of evolutionary biologists from 22 countries, who are all members of their national academies found "no evidence for god."

76 Made by Nature and Man.

77 BioLogos Foundation Meeting, New York, November 2009.

78 In another Christian biology book, *Random Designer* by Richard Colling, an Evangelical author is brave enough to mention Darwin's name, but only once to say that natural selection had been around long before Darwin mentioned it. Karl Giberson, a physicist and Evangelical, in his book *Saving Darwin* does discuss Darwinian evolution and references Darwin many times.

79 In the second through sixth edition, Darwin did add the word "Creator" in his last paragraph. He later called this insertion a mistake in a letter to a friend.

80 *On Seeing Intelligence in Unintelligent Design,* Darrel Falk, The BioLogos Forum, Mar 22, 2010

81 D. Falk, *One Hundred Years ... and Counting*, November 30, 2009, BioLogos Web Site.

82 Falk needs to explain the difference between his book and web site articles on mentioning Darwin.

83 E. O. Wilson, *On Human Nature.*

84 Discovery Institute.

85 Kitzmiller vs. Dover School District.

86 *Design without a Designer*, Proceedings of the National Academy of Science, 104:8567-8573.

87 *Notes on Virginia*, Thomas Jefferson, 1782.

Index

National Association of Biology
Teachers (NABT), 237

National Center for Science
Education, 299

natural selection, (see Darwin's
Theory)

Natural Theology (Paley), 66

Nature Hypothesis, 23

Neuronian Revolution, 39,
95

neuroscience, 95

New Atheism, The (Steger), 16

New Geology, The (Price),
239

Newton, Isaac, 41, 47, 262

Ninety-Five Theses, (Luther),
125

O

Quantum Questions, 101 (Ford),
40

*On the Origin of the Species by
Means of Natural Selection*
(Darwin), xv, 67

Original Sin, 107, 113, 179,
183

Out of Eden (Dawkins), 69

oxytocin, 102

P

Paley, William, *Natural Theology*,
56, 80

parables, Jesus, 98, 111

pedophile crisis, 57, 58

Pius IX (pope), 135

Pius X (pope), 161

Polkinghorne, John, 56,
197

population genetics, 170

Price, George McCready, *New
Geology, The*, 239

Progressive Christians, xiv, 9,
23, 24, 84

Protestant Reformation, 13, 52,
125

R

Randall, Lisa, *Knocking on
Heaven's Door*, 17, 42

Richards, Richard, 10

Richard Feynman, 42,
291

Ruse, Michael, *Can A
Darwinian be a Christian?,*
104

Rushdie, Salman, 188